An Absolute Beginner's Guide to

Keeping a Pet Rabbit

An Absolute Beginner's Guide to
Keeping a Pet
RABBIT

Handling · Feeding · Housing · Grooming

Adriana Cinteza

Storey Publishing

The mission of Storey Publishing is to serve our customers by publishing practical information that encourages personal independence in harmony with the environment.

Edited by Lisa H. Hiley
Art direction and book design by Erin Dawson
Text production by Jennifer Jepson Smith

Cover photography by © Arlee.P/Shutterstock.
com, back t.r.; © Boban Vaiagich/Shutterstock.
com, front b.r.; © Olga Smolina SL/Shutterstock.
com, front l.; © Rosa Jay/Shutterstock.com, back
b.; © NickyPaint/Shutterstock.com, front 2nd fr.
b.r.; © Sasiistock/iStock.com, front t.r.; © Try_my_
best/Shutterstock.com, front 2nd fr. t.r.
Interior photography by © Rick Dahms
Additional credits on page 143

Storey books may be purchased in bulk for business, educational, or promotional use. Special editions or book excerpts can also be created to specification. For details, please contact your local bookseller or the Hachette Book Group Special Markets Department at special.markets@hbgusa.com.

Storey Publishing
210 MASS MoCA Way
North Adams, MA 01247
storey.com

Storey Publishing is an imprint of Workman Publishing, a division of Hachette Book Group, Inc., 1290 Avenue of the Americas, New York, NY 10104. The Storey Publishing name and logo are registered trademarks of Hachette Book Group, Inc.

ISBNs: 978-1-63586-865-4 (paperback);
978-1-63586-866-1 (ebook)

Printed in China by Toppan Leefung Printing Ltd. on paper from responsible sources
10 9 8 7 6 5 4 3 2 1

TLF

Library of Congress Cataloging-in-Publication Data on file

To my niece,
whose love for animals makes
the world brighter and kinder

and

To my mom,
for her unconditional love
and support

CONTENTS

I Love Rabbits!

My love for rabbits started when I was five years old. My mom and I noticed a black-and-white creature hopping around our yard—from its color, clearly an escaped pet rabbit and not a wild one. After it had come around every day for weeks, we finally caught the rabbit, and just a few days later she had a nest of babies! We had no idea what to do, so we visited our local library to check out books about keeping rabbits. We built a big outdoor hutch for the bunnies, and they thrived! Throughout elementary school I acquired a few more rabbits to breed and sold the babies to my local feed store.

In middle school and high school, my priorities changed, so I kept rabbits only on and off. I thought I might take a completely different path, in auto mechanics and car racing, but once I bought my first house, I got the homestead bug and started my own little farm. At first I raised Angora rabbits for their fiber and New Zealand rabbits because their size fascinated me. In 2011 I started raising Holland Lops and fell in love with their personalities and adorable smooshed faces.

As I learned more about how to socialize and care for rabbits, I decided to start my own company, and that's how Blue Clover Rabbitry began! My Instagram handle quickly went viral, and agencies began to license my photos and videos. I taught myself how to build a website and run a business, with no idea that it would become such a success. I now sell rabbits and run a full-time bunny therapy program and a boarding facility for small animals. Blue Clover Rabbitry is continually expanding, and I am deeply thankful for the knowledge I gain every year. I love assisting bunny lovers to educate themselves and build a successful relationship with their pet rabbits.

Bunny Therapy

In our bunny therapy program, community members come to Blue Clover Rabbitry to snuggle baby bunnies, hand-feed them, and experience the joy of watching them hop around. The idea started in an adult family home that my mother owned. I began bringing baby bunnies in for the residents to pet and noticed how calm the bunnies became—they would sit on the residents' laps for an entire hour without becoming fidgety. After that, I remodeled the back of my shop into what I call "Bun Mansion" so that the broader community could enjoy snuggling baby bunnies as well.

After the remodel I shared an invite on Instagram for anyone to come by and see our new place. I expected about 20 people to show up, but the number was more like 400! I knew right then and there that I had something unique, and people wanted to be a part of it. The bunnies at Bun Mansion range from day-old newborns to eight-week-old babies, and it's a win-win situation, bringing joy to humans while our baby bunnies become well socialized and secure.

Bunny therapy is also a great way to educate people interested in adopting a pet rabbit. During the sessions, I'm happy to answer questions to help people decide if a rabbit will be a good fit for their lifestyle. We also offer a mobile bunny therapy service, bringing baby-bunny joy to homes and workplaces. This developed during COVID lockdown, and it was so successful that I kept it going.

Word quickly spread about our bunny therapy and soon companies such as Microsoft and Alaska Airlines—as well as local hospitals, schools, and nursing homes—began hiring us to host sessions for their employees, students, and residents. If this inspires you, consider registering your own pet rabbit as an emotional support animal, which would allow you to bring it to nursing homes or hospitals to bring joy to people.

Why Rabbits Are Great Pets

Rabbits can be a wonderful addition to many households. They are particularly well suited for life in close quarters such as apartments or condos. Although rabbits are considered exotic pets by most veterinary practices, they are not difficult to care for. Their needs differ from those of dogs or cats, however, and it's important to educate yourself before you bring a rabbit home. This book will guide you through the world of pet rabbits—from finding the right rabbit for you to understanding rabbit behavior, feeding and housing your rabbit, and learning about basic bunny healthcare.

You may have already decided that a rabbit is right for you, but in case you're still wondering, here are some of the many reasons rabbits make wonderful pets.

They provide comfort and companionship. Most socialized rabbits can be very affectionate toward humans and often show it by grooming and licking their owners. Rabbits can be registered as emotional support animals.

They will use a litter box. Rabbits like to keep their living areas clean and are typically easy to potty train. As long as you empty out their litter boxes regularly, there should be little to no odor in your home.

They require simple living quarters. Rabbits can be allowed to roam freely at all times, but for most people, a cage with a pen around it is the best setup. Rabbits require regular exercise and cannot be kept confined in a cage all the time.

They are quiet. Rabbits don't bark or meow; in fact, they rarely make audible noises unless severely frightened or hurt.

They are low-maintenance. Apart from long-haired rabbits, most breeds require minimal effort on the part of their owners to keep their fur in good condition. Much like cats, rabbits are clean animals who groom themselves daily.

They are relatively inexpensive to own. After the initial investment in equipment— and the bunny itself!—rabbits don't cost much to feed and maintain. They like to play, but toys can be inexpensive or even free, made from everyday items like toilet paper rolls and cardboard boxes. Rabbits usually have few health issues, though of course these can arise as with any animal.

Are you excited to learn more? Let's hop right in!

CHAPTER 1

What You Should Know Before Getting a Rabbit

IT WAS ONCE COMMON for rabbit owners to purchase their pets cheaply from feed stores or through local advertisements. These rabbits were often kept outside, their hutches rarely cleaned; the owners, especially if children, sometimes forgot about them. Due to a lack of awareness about rabbit behaviors and needs, pet rabbits were often abandoned or resold.

Nowadays, thanks to better information and social media, we have unlimited access to photos and videos of loving rabbit owners who show us what life can be like with pet rabbits inside the home. Rabbits are the third most popular pet in the United States, after dogs and cats, with more than two million of them living with people.

The Basics of Rabbit Care

Rabbits are very intelligent animals who flourish with owners who understand their needs. Rabbits need regular attention and exercise. The ideal setup for rabbits includes a dedicated area, pen, or corner of the house where they have a litter box, food, water, and enrichment toys. I highly recommend providing multiple chew toys to prevent your rabbit from getting bored.

Rabbits are happiest when they can roam freely through the house, but this may not be an option for many owners. While some free-roaming bunnies use their litter boxes perfectly and never chew on anything but their toys, that is not the case for every rabbit. It's a good idea to start your rabbits in a smaller pen and work up to a larger pen or to free roaming.

Rabbits have simple needs and don't cost much to keep. A basic setup includes a comfortable cage with a litter box, a water bottle, and dishes for food.

Feeding

Rabbits have specific nutritional needs. Contrary to popular stereotypes, they can't eat only lettuce and carrots. Their primary diet is hay. Rabbits have an unusual digestive system, with which they eat their food twice! They pass soft droppings called cecotropes, then eat them to absorb more nutrients. This may seem unpleasant, but it is perfectly natural and healthy.

Cleanup

The most important chore for a rabbit owner is keeping the litter box clean. A potty-trained rabbit makes cleanup easy. I use a shop vacuum to do a quick pass for strands of hay and the few stray poops that haven't made it to the litter box. It usually takes me less than five seconds per pen per day. Depending on the litter box and the number of rabbits using it, you may need to empty it every other day, but most need to be cleaned only two to three times per week.

Exercise Needs

Rabbits need at least 30 minutes a day outside of their cage to exercise and bond with you.

It's important for your furry little friends to have time outside their cage each day to exercise. You don't necessarily have to let them roam freely through your entire house, but they need a decent-size area to run and jump. A 36-square-foot pen is a good size for most small to medium-size rabbits to run around in, although more space would be even better. Large rabbits should have about twice that amount of room, as their hops take them much farther.

For all bunnies, plenty of exercise time will enrich their lives and keep their bodies and digestive systems in tip-top shape. It's best to let them exercise while you are home so that you can monitor them while also building your connection with them. Try to give your bunny at least 30 minutes a day of hopping around and/or snuggling with you while you watch TV or read. The more time you spend interacting with your rabbits, the more they will learn to trust you.

Typical Expenses

It's important to consider the costs associated with buying and housing any animal for its entire life. Depending on breed, rabbits can live from 6 to 12 years. Larger breeds typically have shorter lifespans. Here are some estimated costs for purchasing and caring for a rabbit—prices will vary according to your location.

Initial cost. The cost of a rabbit varies widely. I recommend finding a responsible, ethical breeder; the price of a purebred rabbit can range from $200 to $600. Not all breeders are alike and it's important to evaluate them carefully. It is also possible to get rabbits for free or from a shelter, which usually has an adoption fee. See Chapter 2 for more about buying a rabbit.

Housing. An appropriately sized cage or pen, purchased new, can range from $40 to $600, depending on style or materials. If aesthetics matter to you, a pen may cost more. You'll also need a litter box and food and water dishes. See Chapter 5 for more information about housing.

Monthly costs. Ongoing expenses include litter, hay, pelleted food, and maybe some new toys. Plan for a budget of $50 to $90 per month.

Healthcare costs. Unlike cats and dogs, rabbits do not require regular vaccinations—not even for rabies, as they can't carry it. I do recommend an annual checkup to make sure you aren't missing any health issues, especially as your rabbit ages.

Spaying or neutering is a one-time cost; call veterinarians in your area beforehand to check the price. If your local shelter fixes rabbits, that will probably be your cheapest option. Pet insurance, which is fairly inexpensive, can help cover unexpected costs for emergency procedures. See Chapter 6 for more information on rabbit healthcare.

A rabbit can live up to 12 years, so plan your budget accordingly.

Rabbit Anatomy

EARS. Whether their ears are long or short, rabbits have very keen hearing. Ears play a crucial role in regulating a rabbit's body temperature, helping to prevent both heat stroke and hypothermia. The position of a rabbit's ears can also provide valuable insight into its well-being.

Rabbits with floppy ears are more prone to ear issues because their ear canals retain moisture, which can allow bacteria to flourish and make it harder for debris to be removed naturally. It's essential to manually clean the ears of a lop-eared rabbit.

NOSE. Rabbits breathe mostly through their noses and have a distinct way of breathing that involves twitching their noses. A lower rate of twitching indicates that your rabbit is relaxed and feels comfortable in its environment. The external nostrils, or nares, are small and covered in fur.

The nostrils are lined with hairs that keep out dirt and other particles. The Y-shaped inner structure provides efficient air circulation and facilitates a great sense of smell. A wet nose may be a sign that a rabbit is overheating or might be sick.

Rabbits breathe much faster than many other animals. Their typical respiration rate of 30 to 60 breaths per minute can go up to 120 per minute when they detect danger or need to regulate their body temperatures. Their normal heart rate is 130 to 180 beats per minute, which is two to three times faster than a human's. When a rabbit is frightened or distressed, its heart rate can increase to over 300 beats per minute.

EYES. Rabbits are prey animals, so they have large eyes positioned high on the sides of their heads, giving a nearly 360-degree panoramic view of their surroundings. This wide field of vision enables them to spot predators from nearly all directions, including overhead, without moving their heads. In the wild, rabbits are crepuscular, meaning they are most active at dawn and dusk; rabbits do not see well in the dark.

WHISKERS. The whiskers, or vibrissae, help guide rabbits through their surroundings. A whisker is much thicker than a hair and has sensitive nerve endings. Rabbits have whiskers on their cheeks and above their eyes. The longer whiskers help with spatial awareness so a rabbit can avoid getting stuck in small spaces.

FEET. With their large hind feet, rabbits can easily jump six feet or even more, depending on breed, age, and overall fitness. The hind feet move together, propelled by strong muscles in the hind legs. Rabbits often balance on their hind feet to look around or while grooming.

The soles of the feet are cushioned with a protective layer of fur. Rabbits have strong, sharp nails for digging burrows and living underground, and for defending themselves.

TAIL. That adorable, fluffy tail ranges between two and four inches long, depending on breed. The rabbit's tail doesn't have as much functionality or movement as that of some other animals, but it is used to communicate.

The flashing underside of a running rabbit's tail signals "danger" to others. A raised tail indicates that a rabbit is either afraid or feeling aggressive and ready to lunge. Tail raising can also be a sign of submission by a female rabbit who is ready to breed. In warm weather, rabbits raise their tails to lower their body temperature.

Common Rabbit Terms

Binky. A funny movement in which a rabbit jumps in the air and twists its body from side to side. This motion means a rabbit is happy and enjoying open space.

Buck. A male rabbit.

Bunny. A bunny and a rabbit are the same thing; "bunny" is mostly a term of endearment for a baby or young rabbit.

Burrow. An underground tunnel dug by wild rabbits for nesting and hiding. A series of connected burrows is a warren. Domestic rabbits retain this instinct for digging.

Cecotropes. Soft, grapelike droppings, which rabbits usually produce in the early morning or evening. They are high in fiber and are consumed by the rabbit as an important part of their digestive process and gut health.

Colony. A group of wild or domesticated rabbits. Wild European rabbits live in large colonies, but North American rabbits are a different species and typically live independently. (See Pet Bunny, Wild DNA, page 70.)

Doe. A female rabbit.

Fluffle. Another name for a group of rabbits.

Hutch. An enclosed cage for housing rabbits, typically kept outdoors.

Kit. A baby bunny. Mama buns can produce between 1 and 12 kits in a litter depending on size and breed. Usually, the bigger the breed, the larger the litter.

Lagomorph. The animal order that includes rabbits, hares, and pikas. Rabbits are not rodents, despite their large front teeth.

Litter (also nest). A group of baby bunnies born from one mother. A doe can have several litters per year.

Loaf. A rabbit posture in which it lies with its legs tucked under its body, looking somewhat like a loaf of bread.

Thumping. A drumming motion made with the hind feet. Rabbits thump for several reasons: to indicate distress or displeasure, to mark their territory, or to warn others of danger.

Warren. An interconnected system of burrows dug by wild European rabbits for shelter.

Baby Bunny or Older?

You'll have to decide whether you want to start with a baby bunny or if you're open to adopting an older rabbit. One thing to consider is that rabbits can live up to 12 years, depending on breed. Larger breeds tend to have shorter lifespans. There are pros and cons to adopting at any age, so let's talk through them.

Baby Bunny Pros

Socialized babies are usually the easiest to bring into to a new home. Baby bunnies are not yet hormonal and, if raised in a social environment, should be able to adapt well. You will find it easier to teach them social skills, which builds a bond and makes them less skittish, more enjoyable pets. Whether you have a busy house with children or a quiet home, a baby bunny is likely to adapt more easily than an adult rabbit.

The legal age in most states for baby bunnies to be rehomed is eight weeks old. At this age they bond easily because they haven't become territorial yet. Not all baby bunnies are raised equally, however. If you adopt from a source that handled and socialized their baby bunnies, you shouldn't have any issues with bonding. If you adopt from someone who has done little or no socializing, it could take a while for your bunny to build up trust, but it's still easier to form a relationship with a less socialized younger rabbit than an adult one that hasn't been handled very much.

A properly socialized baby bunny should adapt easily to a new situation and form bonds with people.

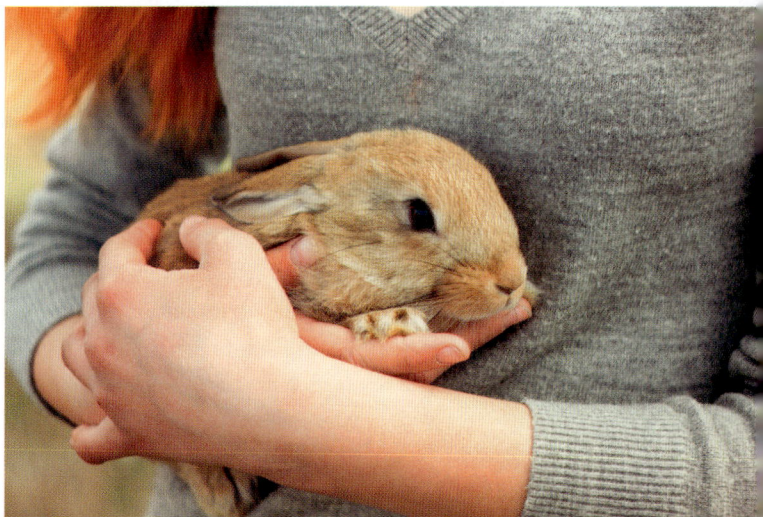

It's also much easier for young rabbits to adapt to a new environment, assuming they haven't had any traumatic experiences in their first eight weeks of life. It's usually easier for a baby to become accustomed to other pets, such as cats and dogs, as long as your other pets do not have a high prey drive and don't do anything to frighten the bunny. If an adult rabbit has not been exposed to other animals, it could be trickier to acclimate it (see Introducing Rabbits to Other Pets, page 87).

Baby Bunny Cons

You may need to litter-train a baby rabbit (see page 113). Most rabbits will pick up good litter box habits quickly if they are provided with the appropriate setup. Although bunnies can learn to use a litter box at eight weeks, the development of sex hormones between three and five months of age will cause males, and occasionally females, to mark their territory by spraying urine. Getting them fixed will correct this behavior, but the typical age for spaying and neutering is about six months, so you may need to be patient for a couple of months. Limiting their space during that time and offering a litter box at least twice their size, or even bigger if it can fit in their area, usually helps litter habits. Once the rabbit is fixed, it can be reintroduced to larger areas or begin to freely roam.

Spending a lot of time with your new baby bunny can help determine how its personality develops.

Baby bunnies' personalities are unformed and can change a lot through their hormonal stage. That means you won't know if your new pet is particularly mellow, for example, or more inclined to be energetic. The way their personalities develop can depend on how their owners interact with them in their developmental stages. If a baby bunny has been socialized daily by the breeder, it should remain social with its new owner, but only if proper handling continues.

This isn't a con, really, but young rabbits have particularly sensitive digestive systems, so it's imperative to feed them properly (see Chapter 4) and make sure they get plenty of exercise to keep their gut moving.

Adult Rabbit Pros

Providing a much-needed home for an adult rabbit can be a great option. Once a rabbit is fully mature, its personality is more defined, and it may be easier to tell whether it fits your situation. You may also be able to find one who is used to other animals if this will be an issue in your home.

Adult rabbits are likely to have already been spayed or neutered, particularly if you adopt from a shelter, where this is usually a requirement.

Most adult rabbits are potty trained, so bringing them into a new home can be easier than it is with younger rabbits; with an older rabbit, you can avoid the months of hormonal adolescence.

Adult Rabbit Cons

Older rabbits are more likely to have developed behavioral issues—after all, there is usually a reason they are being rehomed. Always ask the previous owners why they are seeking a new home for their pet. An unsocialized or very fearful rabbit may have developed a habit of biting in self-defense. Another reason some people rehome their rabbits is that they aren't getting along with another rabbit or other pet in the house. If you have other pets, ask if the rabbit has been around other animals and how it reacted.

The most common reason for rehoming an older rabbit is that the family members, especially young children, have lost interest. If the rabbit was neglected, it could have health or personality issues. It may take some work to rebuild trust and help it feel loved and safe in a new environment (see Adopting from a Previous Owner, page 52).

An older rabbit may be less adaptable to a new environment or lifestyle. If the rabbit came from a home where it lived outside and wasn't socialized, moving into a home with a lot of activity or with young children may be stressful. Noises may trigger it to run and hide, and it could be harder for your bunny to bond until it becomes used to all the new stimuli. It's not impossible to form a good bond with older bunnies, though. There are many things you can do to help them feel comfortable in your home, but only time will tell how well they will adjust.

An adult rabbit can be a great choice for a family, especially if you know something about its background before you bring it home.

Buck or Doe?

If you are looking for a pet rabbit rather than breeding stock, the sex of the rabbit doesn't have to play a big role in your decision. Many people say that bucks (males) make better pets than does (females), but this statement really isn't true. Bucks and does are equally likely to have calm or energetic personalities.

Both sexes can become hormonal as early as eight weeks old (although the shift is more likely to occur at three to five months). Bucks are usually more rambunctious during this stage than does. Getting them spayed or neutered helps reduce hormone levels and typically calms them down. Most veterinarians will not spay or neuter a rabbit until it is five to six months old, so it's good to be prepared for your bunny's adolescence.

Hormonal bucks are also prone to spray their urine to mark their territory; not every buck will feel the need to spray, but it's best to realize that this is a possibility. Does rarely spray but have been known to do so. If you have an issue with your baby bunny not using its litter box, this may not resolve until the rabbit gets fixed.

A male rabbit is called a buck.
A female rabbit is called a doe.

Is It Okay to Have a Single Rabbit?

Rabbits can thrive in pairs or even small groups, but they also can live solo. A single rabbit is best suited to someone who works from home or is at home most days. The more time you spend with your rabbit, the better your relationship will be and the more social your rabbit will become.

If you work outside the house or are not home much of the day, it's nice for your rabbit to have a companion, but this is not a requirement for it to be happy. They are very independent animals.

Adopting a Pair

If you want a bonded pair of rabbits, the easiest way is to adopt two babies younger than three months old. They don't have to be from the same litter or even the same breeder, but the younger they are, the better. They are more likely to form a bond and understand that their territory is shared when hormone levels are lower and both are very new to their new home. The more time a solo bunny has on its own in a new home, the more territorial it will be, which can potentially make bonding more difficult or time-consuming.

Some people choose to adopt two of the same sex to avoid having an accidental litter. Though this seems to be the easy route, two rabbits of the same sex may get into scuffles or fights once their hormone levels start rising. Scuffles are not a big deal, but if they do start aggressively fighting, the pair may need to be separated until they are spayed or neutered. Some same-sex rabbit pairs come bonded as babies, only to have their bond disrupted due to hormone levels. While most will resume their friendship after getting spayed/neutered, be aware that there is a small chance they may not bond again.

If you adopt opposite sexes, you must separate them by 11 weeks of age so that you don't end up with an accidental litter. This means they must be in separate enclosures and cannot come into contact. They can be in the same room or a pen with a solid or double divider that separates them completely. Rabbits can mate with a single layer of gridded divider between them! It's a good idea to recheck the sex of your rabbits when they are around three months old because accidents happen. I can't tell you how many times I've read of sellers claiming that a pair is of the same sex and the buyers being surprised when one of the pair has a litter a few months later. Once your male rabbit is fully developed, between three and four months old, it will be very easy to see testicles through his fur. See page 130 for more about determining the sex of rabbits.

Male rabbits can stay fertile up to one month after they are neutered, so it's best to wait at least a month to introduce or reintroduce your opposite-sex rabbits. Most people choose to have just the male of an opposite-sex pair fixed, as the surgery is less invasive and less expensive, and it solves the issue of marking territory with urine.

Rabbits can be happy living alone or with another rabbit.

If you have had to separate your same-sex rabbits, you can reintroduce them once their hormonal behavior calms down. Every rabbit is different, so this phase could take days or months. Again, after a rabbit gets spayed or neutered, it usually becomes calmer and less territorial.

Introducing a New Companion

What if you have a single rabbit and wish to introduce a new bunny to your home? Rabbits can bond to one another regardless of sex. Rabbits have distinct personalities and preferences, however, and you may have to try a few different bonding strategies. Some rabbits form attachments quickly, and some are very territorial and less inclined to be friendly right away. There are several ways to start the bonding process, and if you do this properly, you are likely to succeed! See Chapter 3 for more on introducing a new bunny.

Are There Hypoallergenic Rabbits?

If you want a rabbit who sheds minimally and is considered "hypoallergenic," a rabbit with rex fur is a great idea. Rex fur is short, dense, and silky soft. Breeds with this type of fur include full-size and Mini Rexes, Mini Plush Lops, and Velveteen Lops. The latter two are recently developed breeds that aren't yet recognized by the American Rabbit Breeders Association (ARBA), but with some research you can find breeders who are close to perfecting them.

I put the word "hypoallergenic" in quotes because no animal is 100 percent hypoallergenic. Rabbits with rex fur, however, shed less and produce much less dander than other breeds. They require very minimal grooming. The texture of their fur is incredibly soft—you almost have to feel it to believe it! I used to raise Mini Plushes and noticed that people who were allergic to Holland Lops sometimes had no reaction to the Mini Plushes.

FULL-SIZE REX

Which Breeds Make the Best Pets?

The American Rabbit Breeders Association (ARBA) recognizes 52 breeds. Determining the most suitable one for you can be a daunting task, but here are a few guidelines to help narrow the options.

Start by considering the space you have to accommodate a pet rabbit. Smaller breeds such as Holland Lops, Netherland Dwarfs, Mini Rexes, and Lionheads can live happily in smaller areas, whereas the largest rabbits, Continental Giants, need much more space and will not thrive in an enclosure that is too small for them.

Your household dynamic is also important to consider. Are there children in your home? Is your space quiet and serene, or filled with the commotion of a family coming and going? The rabbit you choose should be a good fit for your lifestyle.

Next, consider what level of socialization you want in a pet rabbit. Rabbits can be quite friendly, and some breeds seem to have more social tendencies than others, but the handling they receive as babies often plays a more significant role than any inherent predisposition. I have experience with almost every breed of rabbit, and though they all have the capacity to be social, certain breeds do tend to be more independent or reserved in their behavior. A Netherland Dwarf, for example, despite being one of the smallest breeds, typically exhibits a high level of independence; most of the ones I've met have not enjoyed being held. They can be social and come up to you for treats, but this breed is not a cuddly type of bunny compared to Holland Lops or Angoras.

A major part of choosing a breed is also choosing the right breeder. If you are looking for a social rabbit, don't obsess over a random photo of a particular breed you came across on a social media post. Do your research and ask questions! Learn more about finding the right breeder in Chapter 2. In the meantime, pages 29–41 present brief descriptions of some of the more popular breeds.

SOME POPULAR RABBIT BREEDS

American Fuzzy Lop

American Fuzzy Lops have been popular pets since they were recognized by the ARBA in 1989. They have compact bodies and fluffy coats that come in several different colors. They originated as a cross between Holland Lops and Angoras but have coarser fur that needs frequent grooming. They are social rabbits who tolerate or even enjoy being cuddled.

Weight: Up to 4 pounds

Personality: Very social and capable of being snuggly

Grooming Level: High maintenance

SOLID TORTOISESHELL

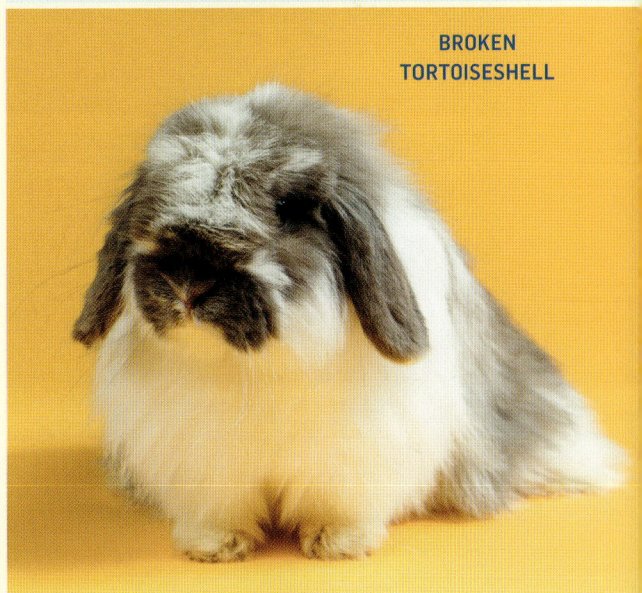

BROKEN TORTOISESHELL

Angora

There are four ARBA-recognized Angora breeds: English, French, Giant, and Satin. Angora rabbits are famous for their long, fluffy wool, which can be spun into yarn. All require daily grooming. Their coats can be sheared every few months to make grooming easier, but the wool can get matted very quickly if not attended to. This breed comes in many different colors. Although high-maintenance in terms of grooming, they are among the more sociable and easy-to-handle breeds.

Weight: Up to 10.5 pounds (Giants can reach 12 pounds)

Personality: Social, docile, easy to handle

Grooming Level: Requires daily attention

FRENCH ANGORA, POINTED WHITE

ENGLISH ANGORA, LILAC TORTOISESHELL

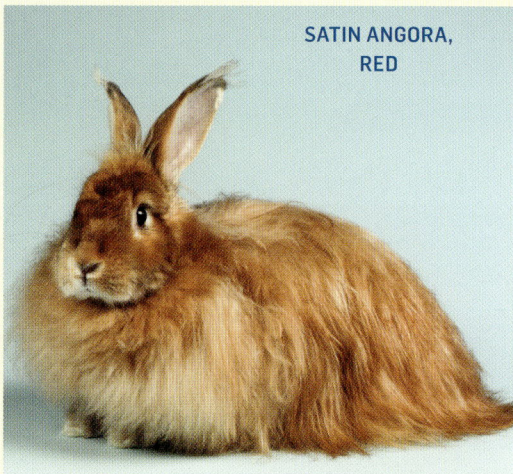

SATIN ANGORA, RED

Champagne d'Argent

Developed in France at least 200 years ago, this breed is prized for its silvery fur with black tips. The face and ears are often dark. Kits are born black and lighten as they grow up. A larger rabbit with a typically calm temperament, a Champagne d'Argent is a good choice for both beginning and experienced owners.

Weight: Up to 12 pounds

Personality: Friendly, independent, capable of being snuggly

Grooming Level: Low maintenance

Dutch

This compactly built breed was developed in England in the 1830s. Dutch rabbits are easy to recognize with their solid coats and distinct white markings that wrap around the neck, shoulders, and belly. It almost looks like they're wearing a vest. Their short fur is often seen in black and white, but several other colors are recognized. Dutch rabbits are low-maintenance and considered very social.

Weight: Up to 5.5 pounds

Personality: Very social and capable of being snuggly

Grooming Level: Low maintenance

TORTOISESHELL AND WHITE

BLACK AND WHITE

Dwarf Hotot

Developed in Germany, the Dwarf Hotot is one of the smallest breeds and is similar to the Netherland Dwarf. It was recognized by the ARBA in 1983. These striking rabbits are pure white with a line of black around their eyes; their fur is dense and soft. Dwarf Hotots are independent but are also capable of being social.

Weight: Up to 3 pounds

Personality: Social and capable of being snuggly

Grooming Level: Low maintenance

English Lop

English lops are thought to be one of the oldest breeds in existence. Their extremely long, floppy ears can span 30 inches from tip to tip. These exceptional ears require diligent care from owners to prevent injury—the rabbit's living area must be free of sharp edges, and doors must be opened and closed carefully when the rabbit is nearby. English Lop fur comes in many colors and is fine, silky, and medium length. This breed is typically calm and easy to handle.

Weight: Up to 12 pounds

Personality: Social, docile, easy to handle

Grooming Level: Low maintenance

TRICOLOR

RUBY-EYED WHITE

Flemish Giant

This breed comes from Belgium, where it was known as early as the sixteenth century. Flemish Giants are massive rabbits who typically weigh between 14 and 20 pounds. They need a much larger living space than smaller breeds. They tend to have a lifespan of only five to seven years. Their dense, medium-length fur comes in a variety of solid colors. They are typically very social and capable of being snuggly.

Weight: Usually 14 to 20 pounds (but they have been known to reach 50 pounds!)

Personality: Very social and snuggly

Grooming Level: Low maintenance

SANDY

Havana

The handsome, compact Havana is one of the oldest breeds recognized by the ARBA. Known as the "mink of the fancy" for their thick, soft fur, they come in five colors: chocolate, black, blue, lilac, and broken (white with markings).

Weight: Up to 6.5 pounds

Personality: Calm, playful, affectionate

Grooming Level: Low maintenance

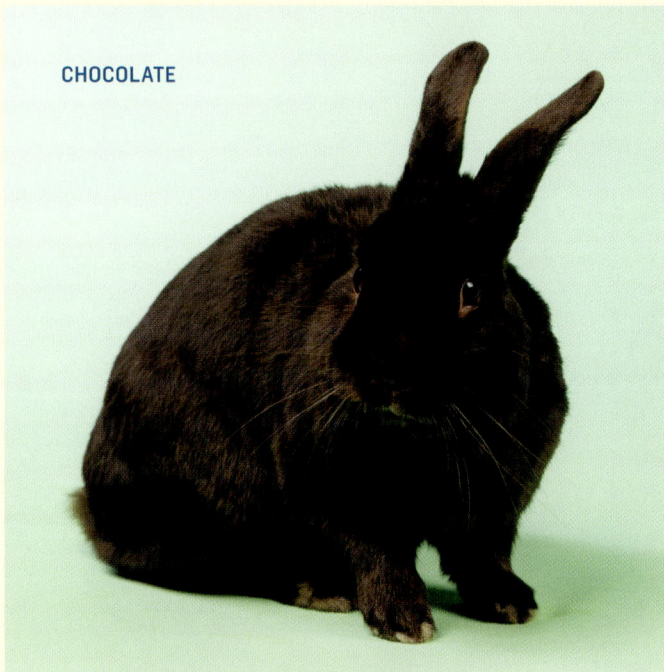

CHOCOLATE

BROKEN BLACK

Holland Lop

Holland Lops originated in the Netherlands and were recognized by the ARBA in 1979. These small bunnies are favorites with many rabbit owners because of their great temperament and small size. They are the smallest breed of the lop-eared family, ranging between 2 and 4 pounds. Their soft fur comes in many different colors.

Weight: Up to 4 pounds

Personality: Very social and capable of being snuggly

Grooming Level: Low maintenance

BROKEN

TORTOISESHELL

Lionhead

Originating in Belgium, these rabbits are named for the distinctive ruff around their heads, reminiscent of a lion's mane. This ruff is about two inches long, while the fur on their backs is shorter. Many Lionheads also have longer hair around their bottoms. They come in several different colors. This breed requires weekly grooming if the fur is kept long. One of the smaller breeds, Lionheads share personality traits with Netherland Dwarfs. They tend to be sassy and independent yet can also be social.

Weight: Up to 3.75 pounds

Personality: Social but less likely to be snuggly

Grooming Level: High maintenance

POINTED WHITE

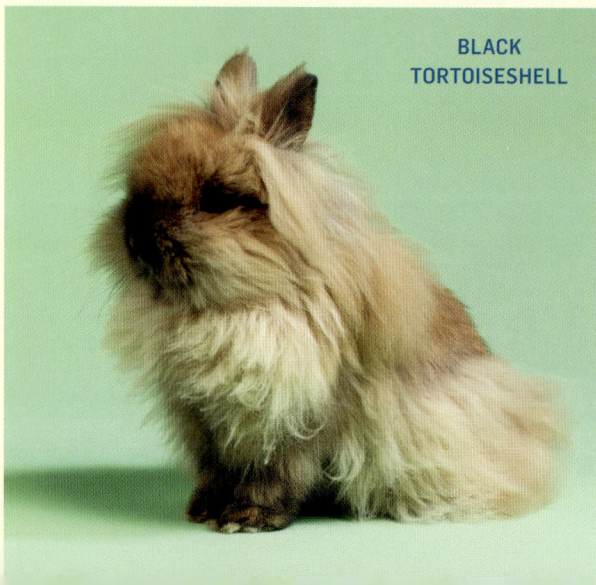

BLACK TORTOISESHELL

Mini Rex

This breed was developed in Texas and was recognized by the ARBA in 1988. The Mini Rex is considered a somewhat hypoallergenic breed due to its short, dense, velvety fur coat. (See Are There Hypoallergenic Rabbits?, page 27.) They shed less than other breeds, and also produce less dander. Adults may weigh as little as 2 pounds, and their coats come in many different colors. They are known for being social pets. The larger Rex breed (up to 10.5 pounds) is also a good pet.

Weight: Up to 4.5 pounds

Personality: Social and capable of being snuggly

Grooming Level: Low maintenance

CASTOR

BROKEN BLACK

Netherland Dwarf

Developed in the Netherlands in the early twentieth century, this breed was recognized by the ARBA in 1969. With its tiny size, round face, and small ears, Netherland Dwarfs are one of the more popular breeds. Their soft, dense fur comes in many different colors. Despite their size, they can be quite sassy and independent, but still can be social if handled properly.

Weight: Up to 2.5 pounds

Personality: Social but less likely to be snuggly

Grooming Level: Low maintenance

CHESTNUT

BLACK OTTER

CHOCOLATE

CHAPTER 2

Bringing Your Rabbit Home

YOU HAVE SEVERAL OPTIONS for bringing a rabbit into your home: buying from a breeder, adopting from a shelter, going to a pet store, or contacting a private owner. Let's discuss those choices and then explore what you need to know about introducing a new pet to your household.

Buying from a Breeder

Health and temperament are what matter most when breeding any animal. It's a great idea to adopt from a responsible breeder if you are looking for a bunny to bring into your family as a baby. Breeders also often have older rabbits who are retired from their breeding program or have been returned. Ethical breeders strive to breed healthy rabbits and will make strict decisions on which animals to keep for breeding stock and which to release only as pets.

Unfortunately there are many unethical breeders who are focused on money and don't care about the welfare of their rabbits or what kind of homes they end up in. Do your research and find a breeder with a solid reputation who will stand behind their animals. Avoid making an impulse purchase based on cute social media posts!

Finding a Responsible Breeder

An ethical breeder selectively breeds and raises animals who will be healthy and make good pets. Rabbits from good breeders are typically social and adaptable, as they've been handled from an early age and desensitized to various sounds and situations. Though it's more expensive to adopt from an ethical breeder, it's worth the investment in the long run. I know of many people who couldn't afford to buy from a responsible breeder, then wound up with a cheaper rabbit and a lot of questions. Breeders who sell more rabbits at lower prices may not be able or willing to offer support once the rabbit leaves their property. It can be hard to find free, reliable information, so it's especially important for first-time rabbit owners to adopt from a knowledgeable, experienced person. If you need to rehome your pet for any reason, the breeder should be able to help you or take the rabbit back.

You are more likely to get a well-socialized pet from an ethical breeder.

Following a rabbitry's social media pages or joining rabbit groups online is a great way to learn more. Read through the posts and see what others have to say. Feel free to ask these groups if there are responsible, ethical breeders in your area. Most forums also tend to have rabbit experts who post information. If you're following a particular breeder, you may notice that people who have adopted from them still follow their posts or comment frequently. You can reach out to those individuals for references.

If a rabbitry has a website, read it thoroughly! In particular, look for a detailed sales policy. It's critical to read the entire policy to see what the breeder stands for and whether they take responsibility for the animals they bring into the world. Look for disclaimers about the health of the rabbit for sale. A good breeder will guarantee the health of their stock at the time you pick up your rabbit. If the rabbit you buy seems unwell when you arrive, the breeder should be willing to either bring it to full health before you take it home, refund you, or let you choose a different bunny. If you have signed a policy with a "sold as is" clause, you have little legal recourse.

It is normal for breeders to state that once an animal leaves their facility, they are no longer responsible for any health issues. This is not a red flag but a way for breeders to protect themselves, as they cannot control how the animals are cared for after they leave the rabbitry. This is why it's important to check the rabbit completely before you leave the property. You can even schedule a veterinary visit for the same day or the day after you adopt a bunny to make sure the rabbit is in good health. An ethical breeder will work with you if a problem is discovered very soon after your adoption.

A responsible breeder will also have a take-back policy that states that if an owner can no longer care for their rabbit for any reason, they can return it to the breeder. If this isn't mentioned specifically, or if the breeder says they will not take back any animals after they've left their property, do not adopt from them! Even if you have no plans to return your rabbit, it's not a good idea to support this breeder. Animal shelters should not be burdened with taking in unwanted animals that people have adopted from breeders. It is the breeder's responsibility, and they should be held accountable for every animal they produce.

Questions to Ask a Breeder

It's important to ask questions to make sure the breeder is the right fit for you. If you can, visit the rabbitry so you can meet the people who work there and see how they care for the rabbits. Many small rabbitries are closed to the public, as the breeders keep their rabbits in their homes. They should, however, be willing to send photos and/or videos of their property and available bunnies to interested people, and to answer questions and explain their policies and practices.

Make sure the breeder is taking good care of the adult rabbits as well as the babies. Every rabbit should have a properly sized cage with room to move around, and get free time outside of the cage for exercise.

Check Things Out!

I once went to look at the website of a popular rabbitry I'd been told about. The pictures showed rabbits in tiny wire cages, and it took ages to scroll down their list of hundreds of available babies. When I called to ask some questions, the owner mentioned that they had more than 120 female rabbits and around 60 males who were bred yearly, but mostly around Christmas and Easter. That alone was a huge red flag, but it got worse.

I asked if the rabbits ever got time out of their cages. After a long pause he said, "They sometimes do when we have time," which made me sure they didn't get outside exercise. Managing a rabbitry with 20 to 30 rabbits is my full-time job, so I asked how many employees he had. He stated it was just him and his mom managing 180 adults and more than a thousand babies per year.

Run from a breeder who has these qualities. Report them to animal control and move on to find a more responsible, ethical breeder. Too often I hear of people "rescuing" animals from bad breeders who don't have clean facilities or are abusing animals. But this only supports these breeders so they can produce more animals.

If housed outdoors, the rabbits should have adequate shelter from wind, rain, and sun. The facility should be clean, and the rabbits should look alert and healthy (see Recognizing a Healthy Rabbit, page 56).

Whether you can go to the breeder's property or not, here are some questions you should ask.

What is your experience with rabbits? How long have you been in business?

More than 90 percent of rabbitries close their doors after the first three years of breeding because it is a labor-intensive job that can be overwhelming, especially for breeders who work another job as well. This doesn't mean you should only buy from long-established rabbitries—everyone has to start somewhere—but the breeder should be able to answer any basic questions you have about rabbit care. Most potential owners have questions about potty training, socialization, and feeding, so it's important to find a breeder who is willing to educate you.

What support do you offer to owners who buy your rabbits? If you close your business, will I still be able to reach out to you for support?

If you are a first-time rabbit owner, you will likely have questions once you bring your new pet home, even if you've done a lot of research. You will receive better and more helpful advice from an established breeder who offers lifetime support to their customers. You'll want to make sure you have their contact information in more than one form. You should be able to ask them questions down the line, even if they are no longer selling rabbits.

How many breeding adults do you have? How many litters do you produce a year?

Smaller-scale rabbitries have the advantage of being able to dedicate quality time to their rabbits, resulting in more socialized rabbits who are better suited as family pets. As a point of comparison, I raise about 200 babies each year. I run my business full-time with two part-time employees and have an ongoing bunny therapy program to help socialize those babies. It's still a lot of work. If someone works a full-time job with no one else helping their rabbitry, I'd be leery about how much care and socializing those babies are getting.

How often do you socialize and spend time with your rabbits?

If you want to adopt a social rabbit, look for a breeder who socializes with their rabbits frequently. Baby bunnies don't need to be held for several hours every day to be social, but they should be handled and snuggled for at least 20 minutes a few times per week to help desensitize them. The younger the rabbit is, the easier this is to do. I use week-old babies in my bunny therapy program because they are easily desensitized and comforted by being held in warm hands and stroked. Other forms of socialization can include petting and hand-feeding. Any kind of gentle interaction helps reduce fear of humans and will build trust.

What can you tell me about this specific rabbit?

You hope the rabbit you are interested in has been socialized in the ways mentioned above. In addition, find out how that rabbit has responded to being handled. Does it stay calm or does it move away from people? Some babies are more prone to run up to the pen when I walk by to feed them or take them out for playtime. You can ask your breeder which ones tend to be more interactive.

How are your rabbits housed and cared for? How often do they come out of their cages?

Although most small-scale breeders won't allow you to visit their facility due to health or privacy concerns, there is no reason why they can't share photos of their setup and cages. Hay scattered all over the floor is not a red flag, but look to make sure the rabbits aren't in tiny cages that are overflowing with poop or urine-soaked litter. Any rabbit housed in a smaller-cage setup should be allotted free-range time either indoors or outside.

Can you provide a health guarantee?

Many people ask this question, but it is rare to find a breeder who offers any kind of health guarantee after an animal has left their care. There are too many variables that breeders can't control, including customers who try to scam them, so breeders have to protect themselves. Some breeders do have a 24- to 48-hour clause that allows you to return a rabbit if it falls ill within that time frame. The clause may also guarantee a full or partial refund, or allow you to exchange the ill rabbit for a healthy one. Some rabbitries, including mine, allow their customers to buy insurance that will offer a refund if their rabbit dies from a genetic issue within 180 days of being adopted. We want to know if there is a genetic issue because our top priority is raising healthy rabbits. Always read the sales policy in full before reserving an animal so that you know exactly what you can expect if something goes wrong shortly after you bring your rabbit home.

Avoid Scammers

A great way to avoid scams is to join rabbit groups online and ask group members where they recommend you adopt a rabbit. Most groups mainly promote shelters, so if you are more inclined to find an ethical breeder, look for an online group that supports responsible breeders as well.

If you have never heard of the rabbitry or rescue, or you don't know anything about the person selling, be sure to watch out for the following warning signs before communicating with them.

Fake social media accounts. Most scammers use fake accounts with little history; scroll through to see how long they've been posting. A new account is not always a warning sign, but few followers or little engagement may indicate a fake account or an untrustworthy breeder. Read reviews and make sure they sound legitimate. You can try sending direct messages to anyone who has left a review, to ask about their experience adopting from that specific person or breeder.

Misleading photos. Check for watermarks on photos. Many scammers steal photos from other accounts and post them as their own. Examine the photos for consistency. If photos have backgrounds that appear to have been taken in different climates or regions, chances are it's a scam. If a breeder refuses to send current photos or videos of their rabbits, stay clear!

Up-front payment demands. Scammers are often pushy about collecting money and may use platforms without fraud protection. Do not send money to anyone you have a bad feeling about. If a breeder tells you they will send photos only after payment, do not proceed!

Adopting from a Shelter

Tens of thousands of rabbits are surrendered to shelters every year. Many people do not properly research the care that rabbits require prior to adopting one, or, once they have their new pet, do not spend the time and effort to bond with it, eventually losing interest. Far too many families buy rabbits for their young children around Christmas and Easter, only to find that their children are unable or unwilling to care for them in the months that follow.

While many animals with emotional and health issues wind up at shelters, it is possible to find a social shelter rabbit. You may be able to find a rabbit who is a great fit for your household. Most shelters allow you to visit with their rabbits to see if any are a good match. I highly recommend a visit to learn more about the personality of a rabbit up for adoption and any other information the shelter can share with you.

Baby bunnies are uncommon at shelters and are usually adopted very quickly. If you prefer not to wait until a shelter has adoptable babies, adopting an older rabbit could be a favorable option—and there are so many rabbits who need a second chance for someone to love them.

Cost typically plays a role in where people look for a pet rabbit. Shelter fees depend on location but are usually less than $100. Shelters typically spay and neuter every animal before it is adopted, which eliminates a significant cost to the adopter.

Adopting from a shelter can work out well if you are willing to be patient as your new pet adjusts to unfamiliar surroundings.

Some Issues to Consider

While there are compelling reasons to adopt from a shelter, there are also potential risks. Shelters usually have little information on the backgrounds of the rabbits coming in: where they came from, why they were surrendered, or how they were previously treated. It may take weeks, months, or even years for an abused or neglected rabbit to learn to trust you. When you visit the shelter to look at a particular rabbit, you will know right away if it is skittish around people. It will hide when you walk in, or hunch in a corner breathing heavily with its eyes bulging.

Another fact to consider is shelter rabbits rarely originate with responsible breeders, who help to rehome rabbits that an owner can't take care of. Instead, shelter rabbits may have originally come from less ethical sources, such as a pet store. Compared with rabbits from reputable breeders, rabbits from shelters have a higher probability of health issues. This information isn't meant to scare you away from shelters but to make you aware of the risks.

Unlike an ethical, knowledgable breeder, a shelter won't be able to offer you support over the rabbit's life. Most shelters are so busy taking in and caring for animals that they often don't have the staff to walk you through questions or concerns after you've adopted an animal. Not every shelter employee has the knowledge of an educated breeder, so you may not get the help you need—or, even worse, you may receive incorrect advice.

Adopting from a shelter can work out very well if you know what issues to be aware of.

Finally, if you are a first-time owner and don't have a reliable resource for your questions on rabbit care, adopting from a shelter may result in a less enjoyable rabbit-ownership experience. However, if you have owned pet rabbits before and have knowledge about their behaviors and their need for socialization, adopting from a shelter is a great idea. There will always be a need for adopting animals from shelters. Visit a few in your area and meet their current residents.

Please Avoid Pet Stores

Though their rabbits are usually less expensive than a responsible breeder's, it's not wise to buy a rabbit from a pet store. Cheaper is not always better: In the long run, you are likely to end up spending more on a rabbit with health issues. If you are a first-time rabbit owner, it is important to note that pet stores may not provide lifetime support as ethical breeders do. Stores may not have any employees who specialize in rabbit care, so are unlikely to be able to answer questions you have after bringing your rabbit home.

Many pet stores are also not very particular about where they get their animals. They often purchase leftover babies from unethical breeders

Many states and municipalities ban pet store sales of puppies, kittens, and rabbits.

and then sell them to any customer who can pay, even if that customer is not fit to care for them. I once met a man in a pet store who had a box with two white rabbits in it. When I asked about them, he said he was there to sell excess rabbits. Three years prior, he bought his daughter two bunnies, thinking they were both boys, but one was female and of course they had babies. They kept breeding, resulting in litters of heavily inbred rabbits who probably had a lot of health issues. The pet store did not ask him any questions; they knew him well, as he came several times a year to sell them babies. I cannot express how irresponsible this is.

Buying animals from pet stores only pushes along more "product," and the store will continue to restock from unscrupulous breeders. By avoiding the purchase of pets from pet stores, we decrease both the stores' profits and the number of rabbits they accept annually from breeders. In time, this ripple effect may reduce the practice of unethical breeding.

Adopting from a Previous Owner

Expecting every person who adopts an animal to keep it for its entire life is unrealistic. You may be thinking that you are the type of person who would never rehome an animal, no matter what. If you can be that person, great! But things happen in life: divorce, death, loss of a job, moving—the list goes on. When people find they need to rehome their animals, it's often better that they seek a more suitable home rather than keeping them in unfit conditions or taking them to a shelter. If you can give that rabbit a second chance at living in an environment where it is properly cared for and loved, by all means explore this option.

If you are considering acquiring a rabbit from a private owner, the first question to ask is why they are seeking a new home for their pet. If the rabbit has aggressive behavior or health issues, take time to assess whether these are things you are okay with. Ask for details on the rabbit's daily routine.

If you can visit the rabbit before you bring it home, you can determine how well it has been socialized. Ideally, it will approach you, or at least not run and try to hide when you come near. Ask to offer it a treat to see if it will come up to you. If you have it in your heart to offer an older, more skittish bun a lifetime home, knowing it may never want to have anything to do with you, that is a wonderful choice. It may still be possible for that rabbit to become comfortable with you, but only time and patience will tell. If you're looking at this option primarily because it's much cheaper than buying a rabbit from a breeder or adopting one from a shelter, please refrain. You may be disappointed, and that's how rabbits get moved around from home to home.

If you decide to move forward with adopting from a previous owner, it's ideal to know who the breeder was; if the owner shares this information, reach out and let that breeder know you are the new owner. Ethical breeders offer lifetime support that follows the rabbit and not just the initial buyer. If you are not familiar with the original breeder or source, be aware of the risks to your rabbit of unknown health issues. It's a good idea to ask the owner if the rabbit has ever experienced any health issues or if they have any veterinary records that they can transfer.

Before Bringing Your Rabbit Home

A well-prepared, safe, and enriching space is crucial for your rabbit's overall health. It's ideal to start with a smaller, dedicated space or room that is specifically for your furry friend. As your bunny grows, becomes potty trained, and gets more comfortable in your home, you can gradually expand its area until it is free-ranging, if that is your desire. It's important to check for the following potential hazards in any rooms where your rabbit will be allowed to roam.

Electrical cords. Although not all rabbits will chew cords, it's better to be safe than sorry! Block off appliances or any electrical cords that your rabbit would have access to (including those that are behind and under furniture). You can also buy chew-proof wraps for electrical cords to keep your pets safe and your possessions undamaged.

Rabbits like to chew, so keep electrical cords and other potential hazards out of reach.

No Nibbling!

The following common houseplants should be kept well out of reach of curious bunnies.

- Aloe vera
- Caladium
- Dieffenbachia
- English ivy
- Ficus (weeping fig)
- Jade plant
- Oleander
- Peace lily
- Philodendron
- Pothos
- Sago palm
- Snake plant

If you let your rabbit explore outside, keep these plants out of reach as well.

- Alliums (includes onions, garlic, and shallots)
- Bluebell
- Chrysanthemum
- Hemlock
- Hydrangeas
- Lilies
- Lupine
- Nightshades (tomatoes, potatoes, and eggplants, as well as bittersweet nightshade and deadly nightshade itself)
- Poppies
- Rhododendrons
- Rhubarb

Houseplants. Many common houseplants can cause rabbit gastrointestinal (GI) tract issues, diarrhea, or even death. Luckily, rabbits need to consume quite a bit of such plants to get sick. If your rabbit nibbles a few bites of a toxic plant, chances are it will be okay—but it's best not to run the risk. Place all houseplants out of your rabbit's reach and remember that rabbits are clever animals, capable of squeezing through small openings, hopping up on furniture, and climbing to get what they want. Plants on a shelf or windowsill behind a couch are not safe—your rabbit will surely figure out how to reach them!

Escape routes. Rabbits can be skilled escape artists, so whether their area is a solid floor cage or a pen, it should either have a closed top or have walls high enough that they can't jump out of it. Rabbits have been known to clear 3-foot pens, especially if there are castles or other structures inside their pens that they can hop on top of. The safest bet is a 3½- to 4-foot-tall pen. If your rabbit keeps escaping, you can invest in a small camera to watch it when you're not around and learn how it is making the great escape so you can improve its area. If you let your new pet roam, be sure that windows are closed or securely screened. A curious bunny could hop up on a couch or other furniture near a window and fall through. Close doors to any rooms you don't want a rabbit to enter.

Supplies and Equipment

When you adopt your rabbit, ask what brand of pellets it has been fed. There's nothing to worry about if you choose to transition your rabbit to a higher quality or different brand, but it's a good idea to ask for a bag of the pellets your rabbit is used to so you can gradually switch the diet. Most rabbits transition easily and quickly (see Adjusting to New Foods, page 60).

Rabbits need mental and physical stimulation, and they need to chew to keep their teeth in good shape, so have a few different toys ready. Simple toys can be made from toilet paper rolls, paper bags, cardboard boxes, and newspaper. You can collect small twigs from nontoxic trees such as apple, willow, or pine for bunnies' enjoyment and dental health. Several companies specialize in inexpensive rabbit toys. As you learn about your rabbit's personality, you will figure out what toys it is drawn to and be able to bring home its favorites.

In addition to having food and supplies ready, make sure your rabbit's pen or cage is set up and ready for it to explore and relax in. Every rabbit adjusts differently, so it's good to have the area prepared before your bunny arrives, especially if your rabbit is likely to feel timid in the new environment. Rabbits also enjoy structures where they can hide or rest, such as tunnels and small pet castles. See Chapters 4 and 5 for more detailed discussions of feeding and housing your rabbit.

Simple toys can be made from toilet paper rolls, paper bags, cardboard boxes, and newspaper.

Recognizing a Healthy Rabbit

It is crucial to make sure your new rabbit is healthy prior to bringing it home. If you have the opportunity to pick up your rabbit in person, look it over to make sure it is in good condition. If someone else is transporting your rabbit to you, there's nothing wrong with asking the breeder or seller to first send you current, close-up photos of the rabbit to make sure it looks healthy. Here are some specific things to check. For a more detailed discussion of rabbit healthcare, see Chapter 6.

EYES. A healthy rabbit has shiny eyes. A little crud in the corner is normal, but the eyelids shouldn't look puffy and red, or appear wet.

NOSE. The nostrils should be clear of any discharge or excessive wetness. Wetness is different than thick mucus, which usually indicates an infection. Check for scabbing around the nose, as that can be a sign of treponematosis, or rabbit syphilis (see page 136). Both conditions are curable, but I recommend having the breeder bring the animal back to full health before you adopt.

VENT AREA. Look at the rabbit's vent area, where its genitals are. Is it clean? (A little yellow staining is okay.) Check for scabbing, as rabbit syphilis is fairly common. When it flares up, scabs appear on the mouth, nose, and/or vent area. Syphilis is easily cured, and the breeder should treat the rabbit before releasing it to you.

FEET. Look for fluffy, furry feet with no open sores or crooked toes. Some yellow fur on the bottom of the feet is okay, but heavy staining may indicate urine scalding. This occurs when the rabbit has spent too much time in a urine-soaked area, an indication of improper care.

CLEAR, BRIGHT EYES

CLEAN NOSTRILS

FUR. Scan the entire body to make sure the coat is soft with no bald spots. Bald spots usually indicate mites, which are easily treatable. Breeders should check for this, but mites can sometimes appear months after the rabbit is adopted.

BODY CONDITION. Make sure your rabbit is an appropriate weight for its age and breed. Its backbone should not protrude, and its sides should be firm and full, not sunken, with visible ribs. When you pick up the rabbit, it should have some substance.

TEETH. A rabbit's teeth should be straight, with the top teeth and bottom teeth in alignment and not overgrown.

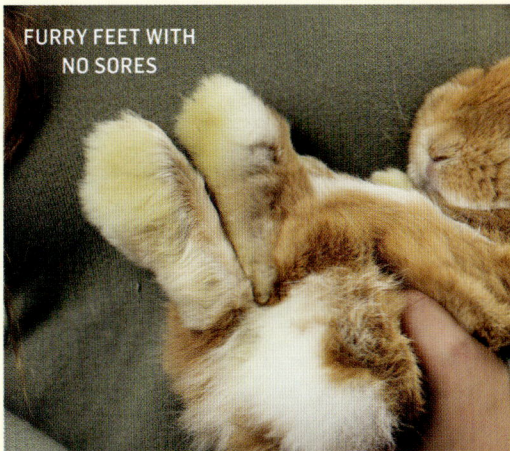

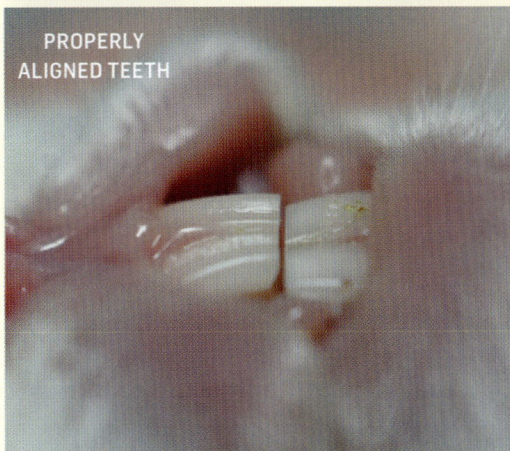

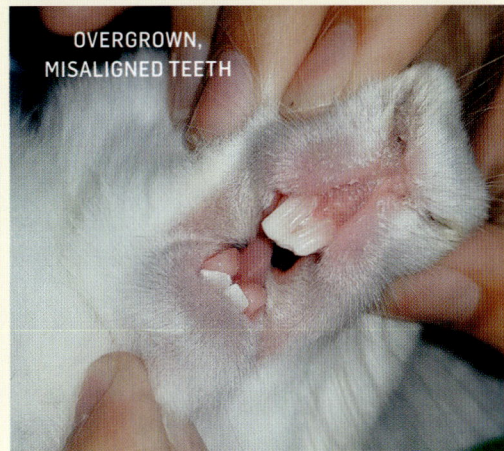

FURRY FEET WITH NO SORES

HEALTHY BODY CONDITION

PROPERLY ALIGNED TEETH

OVERGROWN, MISALIGNED TEETH

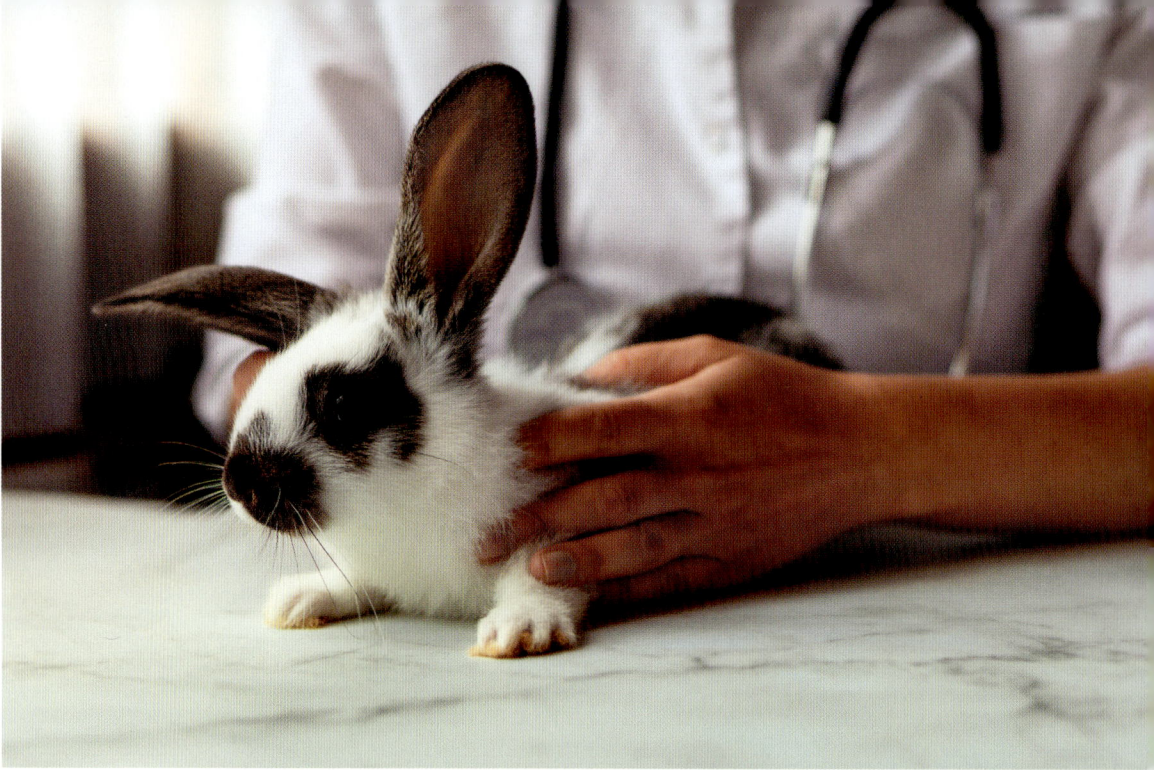

Find a Good Veterinarian

Rabbits typically have few health problems, but it's a good idea to take them for regular checkups by a veterinarian who specializes in exotic animals.

Although most rabbits have relatively few health issues, problems may arise with their skin, ears, or digestion. Veterinary practices classify rabbits as exotic animals, so before adopting, research veterinarians in your area to find one who can see your rabbit for yearly checkups and in case of emergency. The last thing you want is to be scrambling to find a veterinarian when you have a sick rabbit. Start by asking your breeder for recommendations. Even if they live in another area, they might know good veterinarians where you live.

Another way to find a good veterinarian is to join local rabbit groups on social media and ask for recommendations. Prowl through the comments and see if particular veterinary practices are recommended frequently—this is usually a good sign!

You can also ask your local animal shelter for recommendations. Most shelters spay and neuter all the animals they adopt out, but don't have clinics for other types of appointments. Having a shelter spay or neuter your rabbit is fine, but beyond that you'll want to establish a relationship with an exotic-animal veterinary practice.

Even if you find a veterinarian you like, it's good to know which other practices in the area will see rabbits. Check their hours, as some animal clinics only have one or two veterinarians who work with exotics, and they may not be available every day. You should also try to find a 24-hour exotic-animal veterinary practice for after-hour emergencies. It's best to be prepared and have these numbers on hand.

Making the Transition

Helping your rabbit feel safe and comfortable in its new home should be your top priority. Keep your rabbit confined in a small area at first so that it can learn where its food, water, and litter box are. Handle your rabbit often and spend time with it to help create a bond. After a few days or weeks, begin to let your pet explore to the extent that you are comfortable, or bring it into other parts of your home or yard for outings.

Rabbits have different temperaments, and depending on their background, it may take some time for them to adjust. As prey animals, rabbits are naturally timid, although their personalities vary. A socialized rabbit usually becomes comfortable within the first week. If you have a confident or well-socialized rabbit, instead of one acting shy and reserved, continue socializing with it as it adjusts to its new space. It's best to let a rabbit sit in your lap while petting it firmly from the top of the head all along its back. Repeat this for several minutes or as long as you desire and the rabbit allows it. A social rabbit will enjoy being petted this way;

Once a rabbit feels comfortable in its new home, it will likely come up to you for pets and treats or even jump in your lap to snuggle. A free-range rabbit may run to greet you when you come through the door, just like a dog!

Having a snug place to hide will help your new bunny feel secure in its surroundings.

it shows that the rabbit is submissive and trusting, and this kind of inter-action helps it form a bond with you.

If a new bunny is nervous or scared, give it space while it adapts to its new home. Put it in a quiet part of the house where it won't be upset by noisy children or a barking dog. Set up its pen with several hideouts such as tunnels, cardboard boxes, hay huts, or bunny castles. This will help your pet feel safe while it adjusts to its surroundings and new routine. At first, your rabbit will most likely spend much of its time hiding. If this is the case, leave it mostly alone for several days to a few weeks. When you approach the pen to fill its water and food dishes, move slowly and quietly so you don't spook your rabbit. Never force an anxious rabbit to cuddle with you; this will not build trust. For more information, see Socializing a Skittish Rabbit on page 74.

Adjusting to New Foods

Carefully transitioning your new rabbit from its previous diet is just as important as making it feel comfortable and safe. Rabbits' GI tracts are very sensitive, so you should transition your new rabbit slowly from one brand of pellets to another. If you're introducing your pet to greens it hasn't had before, do this slowly as well. If you have a rabbit younger than six months old, you can offer it very small portions of safe herbs and greens, but be careful not to overdo it. See Chapter 4 for a detailed discussion of rabbit-safe greens.

It's wise to offer only one new vegetable or fruit per week. If your rabbit gets sick after trying many new foods at the same time, it's hard to tell which food was the culprit. Watch your rabbit's behavior and be on the lookout for a difference in its stool. Bunny poop can tell you a lot about what's going on inside. Many new rabbit owners confuse cecotropes, the normal soft droppings that rabbits ingest, with diarrhea. Baby bunnies don't eat their cecotropes as often or as quickly as adult rabbits do. If you find uneaten cecotropes, just move them into the litter box or throw them away so your bunny doesn't step on them and make a mess.

Some rabbits can get very nervous when moving to a new environment, which can disrupt their digestive systems and cause them to have softer poops. If this happens, don't fret. Just give your rabbit some space and quiet to adjust to its new home and make sure it's eating hay and drinking water. In a few days to a week, it should be pooping normally again. However, if it doesn't get better after a week, a trip to the veterinarian might be wise in case something else is going on.

Adjusting to a new home can upset a rabbit's digestion, but it will usually settle down in a few days.

CHAPTER 3

Understanding and Handling Your Rabbit

THE MORE WE KNOW ABOUT RABBIT BEHAVIORS, the more we can understand their needs, which will benefit not only them but us as well. If you can figure out why your rabbit is not using its litter box, why it's chewing inappropriate items, or why it leaps out of your arms whenever you try to hold it, you have a better chance of correcting the behavior or making necessary changes so you can enjoy time with your pet.

Basic Body Language

As with all animals, a rabbit's posture and how it moves can show how it feels and what it wants. For example, you can tell a lot from how it holds its ears. Rabbit ears come in all shapes and sizes, from the tiny pointy ears of a Netherland Dwarf to extremely long ears that drag on the floor, such as those of an English Lop. The smaller the ears, the more "ear control"—or ability to move its ears independently in different directions—the rabbit has. On the other hand, some lop-eared bunnies, due to their head structure and ear length, can't lift their ears at all. In rabbits who can, lifted ears usually mean they are very alert, sensing danger, or just curious.

When a rabbit's ears are in a flattened position or pointed backward, it may be feeling aggressive, angry, or afraid—but it could also be calm and relaxed. Other body language will tell you the difference. If the rabbit's eyes are half closed and it is breathing normally, it is most likely relaxed; bulging eyes and rapid breathing signal anxiety.

If you watch an active rabbit, you will notice that its nose twitches rapidly and almost nonstop. The speed of its nose-twitching corresponds with its respiration and its mood. At rest, rabbits breathe more slowly, and their noses twitch more slowly as well.

AWAKE AND ALERT

Using Their Feet

When a rabbit feels afraid or angry, it may thump its hind feet as a warning. It may also thump a foot to warn other rabbits of potential danger or to indicate displeasure. Rabbits may kick their feet hard to try to get away from danger or to protect themselves.

Loafing Around

If your rabbit is "loafed," that is, sitting with its feet tucked in and looking like a loaf of bread, it is comfortable and resting. You'll notice it breathing at a normal rate, and its eyes may be opened normally, squinted, or shut.

Another position you can find your rabbit in is "frog pose." This is when a rabbit lies on its stomach with its front feet forward and its back feet stretched out long behind it. If you catch your rabbit in this pose, chances are it is comfortable and relaxed.

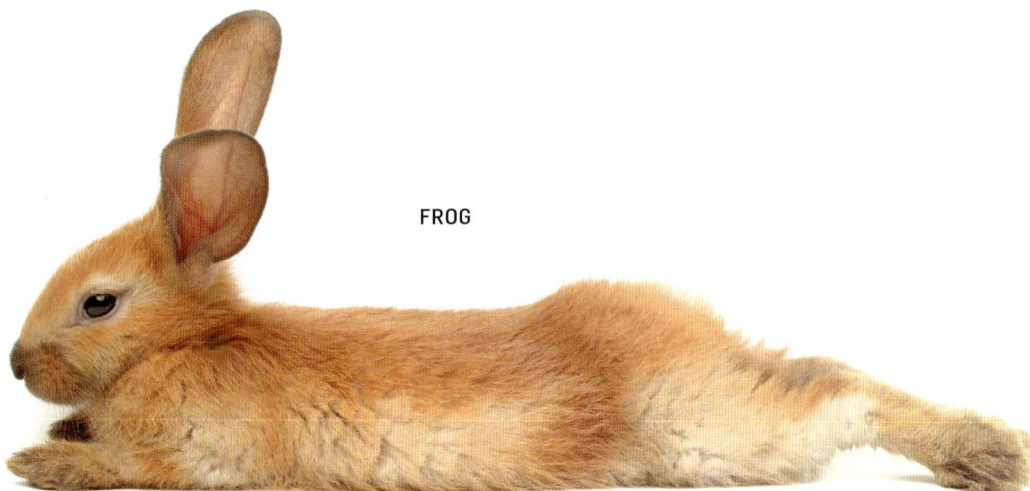

LOAF

FROG

CHECKING THINGS OUT

Upright and Alert

When rabbits aren't resting, they spend time foraging, exploring their space, or playing. The more room they have, the more active they are likely to be. If they are penned up most of the day, without room to run around and play, they will probably lounge around more. A curious rabbit may stand tall on all four feet or sit back on its haunches to survey its surroundings, much like a prairie dog.

Let Sleeping Rabbits Lie

Rabbits sleep in many different positions, most of which are adorable. They often sleep in the loafed position or lying curled on their sides. A pair of bonded rabbits may sleep snuggled side by side or even intertwined.

Sometimes, though, rabbits sleep flat on their sides in what we call the "death position" because it looks like they are not alive. When in this position, they are usually deeply asleep and will not respond to sound or movement around them. Don't panic—just check to see if they're breathing normally, instead of spooking them out of a deep sleep!

Some Common Behaviors

Rabbits are crepuscular, meaning they are most active at dusk and dawn. Because they are not nocturnal, you should be able to sleep through the night without being disturbed if you house them in or near your bedroom. During waking hours, here are some behaviors to observe.

Playing

While rabbits lounge most of the day, they also like to explore and hop around in large, open spaces. Often while doing this, they'll jump in the air, shaking their heads, kicking out their legs, and twisting their bodies. This adorable, fun-to-watch move is called a binky. A rabbit doing binkies is happy, relaxed, and not afraid.

Rabbits also love to investigate their surroundings and play with toys, which provide enrichment and, we hope, deter them from chewing on household items or furniture. In addition to chewing on toys, many bunnies grab them and toss them around. Hay-based toys are a great option, as they provide an extra way for your rabbit to consume hay, which should make up the largest portion of its diet. Castles and tunnels provide areas for them to rest or hide.

Watching your pet "do binkies" (energetically hopping, jumping, twisting, and leaping) is one of the great joys of living with a rabbit.

SELF GROOMING

GROOMING

Grooming

When a rabbit isn't zooming around the room or resting, you will often find it grooming itself. Rabbits groom frequently, so don't be alarmed if you see this behavior throughout the day. A rabbit may scratch its ears with its feet or lick its front paws and use them to wipe its face. Lop-eared rabbits often lick their paws and grab their ears to clean them. They groom by licking their sides and chest as well.

Bonded rabbits often practice mutual grooming by licking each other's forehead. When a rabbit nudges you, it usually wants to be petted on the head, which provides the same satisfaction as another bunny grooming it. Being licked or nudged by a bunny is a wonderful sign that it trusts you and wants to interact.

Chinning

Rabbits have scent glands under their chins that emit an odor only they can detect. They rub those glands on things to mark their territories or to claim an object as theirs. You will never smell any kind of odor once your rabbit has chinned something. If your rabbit chins you, feel honored that it is claiming you as its own.

Chewing

Chewing is the behavior most rabbit owners have an issue with. It's important to understand that this is a natural behavior for rabbits; in fact, it's essential, because their teeth grow continuously and chewing helps grind them down, keeping them in tip-top shape. They are curious

CHINNING

CHEWING

animals, and their mouths are one of the things they use to explore their surroundings. Chewing brings enrichment and mental stimulation. Wooden toys and cardboard boxes are great for chewing and keeping teeth in good condition.

While some free-range rabbits only chew on their toys, leaving furniture and baseboards alone, some rabbits are excessive chewers. A rabbit who chews on its cage or on furniture and walls may be bored or stressed, or need more socialization or space to roam around. Providing more and different chew toys may help redirect this behavior. A rabbit confined in a too-small cage or pen is almost certain to be bored and stressed, so be sure you provide adequate living quarters and opportunity to exercise.

Give your rabbit plenty of space and time out of its pen so it can run freely and explore. Supervise the playtime until you're sure you can trust your bunny not to chew on household items. If left out of its pen for too long, a curious or bored rabbit may start to chew on things other than approved toys. It may take a while to find the right balance of free-roam time, but typically one to three hours per day is adequate for a rabbit who also has a reasonably spacious enclosed pen area. See Chapter 5 for more about housing and play areas.

If your rabbit is doing something that displeases you, never yell or physically correct it in a negative way. Rabbits can become less social or revert to their natural fight-or-flight responses when they no longer feel safe in their environment. Instead, circumvent the issue by rearranging the area or limiting which rooms your rabbit can enter.

One to three hours of supervised playtime a day is enough for most bunnies.

Pet Bunny, Wild DNA

It might help to understand your pet rabbit if you know that all domestic rabbits are descended from European wild rabbits (*Oryctolagus cuniculus*), the kind who dig large underground warrens to live in and raise their kits. The wild rabbits found in North America are a different species, *Sylvilagus floridanus*. They don't dig tunnels but instead scrape out shallow nests in meadows. Females do dig small burrows where they have their babies.

I once had Continental Giant rabbits, which are the biggest rabbits in the world. In the summer, weather permitting, I let them spend most of the day outside in a huge pen with toys and a big castle for shelter. After a few weeks of summer fun, I went to move their pen to a fresher grass area. When I lifted the castle, I discovered that they had dug the longest tunnel I have ever seen. We couldn't even measure it accurately, but it extended more than six feet!

Domestic rabbits descend from European wild rabbits, not North American cottontails. It is in their DNA to dig underground.

Digging

Digging is another natural and unavoidable behavior. If you allow your bunny to roam unsupervised on carpeted areas, be aware that it might start digging and tear up the soft pile, which can quickly make a "carpet spaghetti" mess all over the room! A rabbit typically won't dig at hardwood flooring, so I recommend restricting free-ranging pets to areas without carpeting as much as possible. The shorter the pile, the less damage is likely, but shag rugs and deep pile are a playground for most rabbits. Not only is this rough on your carpets, it's also bad for bunnies to ingest the fibers.

If you allow your rabbit to have any outdoor time, be sure to monitor it—if you end up with a digger, it can make a surprisingly deep hole in no time.

How to Handle Your Socialized Rabbit

Many people struggle with picking up and holding a rabbit because they may not understand rabbit behavior or know how to approach their pet properly. Rabbits are not particularly snuggly by nature, so they don't always love to be held, but socialized bunnies who are bonded to their owners are usually easy to handle with some practice.

Quick, like a Bunny!

Move quickly when picking up your rabbit. If you try to creep up on it or reach out slowly, your rabbit may hop away each time you attempt to handle it. This is a bad habit to allow. When a family tells me that the baby bunny they adopted was easy to handle during the first few days or weeks but then became difficult to catch, I ask them to describe how the change came about. Frequently, the explanation is that they didn't want to scare the bunny, so they moved very slowly when picking it up. If the bunny hopped away, they assumed it didn't want to be held and let it go.

I can't stress enough that you should *not* allow this to happen. Rabbits are very smart animals and will test the limits of what their owners will let them get away with. Each time you go to pick up your rabbit and it hops away, the rabbit feels dominant. It learns that it can hop away without being pursued. Reinforcing the behavior makes it harder to maintain the rabbit's socialization.

The most important tip here is to assert your dominance over your bunny. This doesn't mean being forceful or making your rabbit uncomfortable. The key is to move quickly and confidently when you pick it up.

Scoop your rabbit up with one hand under its bum, to support its back, and one hand under its front legs. It's important to pick rabbits up quickly because once their feet are off the ground, they will feel less secure and may start kicking. Bring the rabbit to your chest, with all four feet toward you, or to your lap with all feet down. You want the feet to be touching something—this makes the rabbit feel comforted and secure. People often try to hold rabbits with feet facing outward, maybe to avoid being scratched, but this position may be unsettling for a rabbit,

Position your rabbit parallel to your body.

Place one hand under its front legs.

Place your other hand under its haunches and hold the back legs.

Scoop the rabbit up with its feet facing you and hold it securely against your body.

Wrapping the rabbit in a blanket can help make it feel more secure while you are learning the correct technique for picking it up.

particularly once it's no longer on ground level, and it may struggle or twist to get away. Practice holding your bunny while sitting on a chair or couch; this is safer for the bunny in case it flails out of your arms.

If you are nervous about being scratched or not picking your rabbit up correctly, you can wrap it in a small towel or blanket so it feels secure and you feel safer. Hold the blanket open with both hands and quickly drape it over the rabbit's back and hindquarters as you scoop it up, cradling it in the blanket. Finish by wrapping the blanket around the feet so that if the rabbit kicks, you are less likely to get scratched.

Once you have your rabbit against you, pet its head firmly to make it feel comfortable and to show dominance. If your rabbit feels safe and secure in your arms, it may spend several minutes or even hours snuggling with you; this depends on the individual rabbit.

If a socialized rabbit is struggling to get out of your arms, it is probably because it does not feel secure. If your rabbit kicks in midair, don't put it back down; just quickly place all four feet against your body, then give it a gentle squeeze to help it feel secure and prevent it from escaping. In most cases a socialized rabbit is not asking to be put back down but just wants to feel stable again. If you put it down, it will quickly learn that struggling results in being let go, which rewards that undesirable behavior.

Socializing a Skittish Rabbit

With a rabbit who isn't properly socialized, you need to be very patient and understand that it may never want to be picked up or held. A great way to help rabbits bond to you is by hand-feeding. Set up a penned-in area around the cage that is large enough for you to lie down in while allowing your rabbit to have some space. Bringing yourself to the rabbit's level shows that you aren't a threat. Start by lying quietly on the floor with some special treats near you. Talk to your rabbit and let it approach the food, but keep yourself still, as any movement could spook it. Don't be concerned if it runs or hides in a corner—that is a natural reaction for a rabbit who isn't used to human interaction.

Once your rabbit feels more confident, you can set up a larger area of about 16 to 20 feet square with pet fencing, and let it explore while you sit in the pen doing a quiet activity: reading, checking your phone, or working on the computer. The point is to let your bunny become comfortable being near you. Once you've done this for a period of time, your new friend may stop hiding and feel safe enough to explore around you, or even come up to sniff you.

Don't reach out to your rabbit at this point, as it will most likely run and hide again. As the rabbit becomes more comfortable with you in its pen, try offering some tasty greens in your hand or placing them next to

Most rabbits will become calmer and friendlier if you are quiet and patient with them.

you so the rabbit can approach for a snack. The next step is to sit upright with your ankles crossed: Put some greens between your legs and see if your bunny will prop its front paws on your legs to grab the greens. Do not pet it yet! This is great progress. Some bunnies never make it this far, but I encourage you to be persistent.

It may take a while for your rabbit to take food from your hand, but this is the best place to start with socializing and bonding. Once your rabbit will approach you and eat in front of you, you can slowly start trying to pet it. If it backs away, that's normal. Once your rabbit begins to take food from your hand, gently swipe the back of your hand over its head as it's grabbing the treat. Again, take things slowly. Over time, the rabbit may trust you to pet it. If this happens, it's a huge step! Eventually, you can try to pick up the rabbit as described in the previous section.

Be patient and respect your rabbit's boundaries. Progress may seem slow, but in the long run, your efforts will speed up the bonding process and make your new pet trust you sooner.

Catching a Nervous Bunny

If you need to take your skittish rabbit to the veterinarian, try putting a travel carrier in its pen with some favorite treats or greens pushed to the very back of the carrier. Wait for the rabbit to hop inside for a snack, then close the carrier door and proceed. Most rabbits are anxious traveling in a car, so put a light towel or blanket over the carrier to make your rabbit feel safely hidden.

If you have a free-roaming skittish rabbit, you may need to block off sections of its living area to get it into the carrier. Your rabbit won't like this experience, but if it's a for a necessary appointment, you have no choice. When you get home again, give your rabbit some delicious treats and let it have some quiet time.

It's a good idea to get your rabbit used to being in a travel carrier.

Rabbits and Children

Having a pet rabbit can be a wonderful learning experience for a child, as long as there is adult supervision. Most children younger than 10 aren't ready for the full responsibility of caring for a pet, although they can certainly help with care and feeding. All children should be taught to properly handle a rabbit and should not be allowed to do so unsupervised until you are sure they and the rabbit will be safe. Depending on your child's age and grasp of proper bunny handling, it may be best for you to hold the rabbit while letting your child pet it.

Before you teach children to hold a rabbit independently, make sure they understand what "gentle" means and can sit still and be quiet when holding the bunny. Start by having them sit on the floor so the bunny can't get hurt if it jumps away. If they aren't wearing long pants, put a towel or blanket over their legs to protect them from scratches, then put the bunny in their lap. Keep your hands close by and help your child pet the rabbit gently.

Eventually you can teach your child how to lift the rabbit and hold it, with one hand under its bottom, one hand around its back, and all four of the rabbit's feet against their chest. Explain that they need a firm hold on the rabbit so it feels more secure and is less likely to jump away, but that they shouldn't be squeezing so tightly that the bunny is harmed or frightened.

In general, children younger than six should stay seated while holding a rabbit. Otherwise, the rabbit could easily flail out of the child's arms and injure its back in the fall. Every child is different, so use your best judgment to decide if your child is old enough or responsible enough to stand while holding a rabbit.

It's safest for young children to sit on the floor while holding and petting a rabbit.

Normal Hormonal Behavior

As mentioned before, rabbits of both sexes go through a hormonal period in which they start acting like teenagers. This stage can start as early as eight weeks of age, but most hormonal behaviors appear between three and five months. Behavior changes might include making humming noises, hopping around your feet, or mounting various objects. Bucks often begin to spray urine to mark their territory, even if they have been using a litter box previously. (A doe may spray, but this isn't common behavior.) They can spray up to eight feet, so you might choose to confine a hormonal buck to a smaller space with solid walls while he goes through this stage. Not every buck will spray urine during their hormonal period, but it's better to plan for it and expect it so you are prepared.

During this time rabbits are more apt to jump out of your arms, even if they were comfortable being picked up in the past. Don't let your bunny get away with it. Not only will it get the wrong idea, but it could fall and hurt itself. Be firmer when picking it up and hold it snug against you while petting its head. You don't need to be forceful, but do show that you are still in charge. Over time, especially if it is neutered or spayed at around six months old, the rabbit will show less aggression and accept your dominance.

Don't be startled if a young rabbit pees on you when you pick it up! Like spraying, this is a territorial behavior that means the bunny is marking you as its property. Often, they choose just one or two people to favor with this attention. Once they are neutered or spayed, the behavior will stop.

What to Do If Your Bunny Bites

There are several reasons a rabbit might bite or nip. If this happens, assess the situation to try and understand why. Where did the incident take place? Were you in your rabbit's territory or in a neutral area? Rabbits can be very territorial, especially when they are not spayed or neutered, so if a rabbit is charging at you or biting, avoid reaching your hand into its pen or near its food. If your sweet baby bunny becomes aggressive after a couple of months, it is most likely a result of rising hormone levels. The behavior will subside after the rabbit is fixed.

There is a difference between biting and nipping. A bite is harder and usually stems from aggression or fear. A nip is much softer but can still sting. If a bite breaks the skin, wash the area with soap and warm water.

If your rabbit is enjoying being petted and you stop before it wants you to, it might give you a little nip telling you to keep going. In this scenario, immediately put the rabbit back on the ground or get up and move away, to make it clear that biting ends the interaction. If you continue petting after being nipped, the rabbit learns that using its teeth will get you to start petting it again. This is positive reinforcement and should be avoided. When a rabbit exhibits unwelcome behavior, never give in to what it wants. Not all rabbits nip to get attention—some will nudge you with their noses instead, which is acceptable and also adorable.

Another way to discourage your rabbit from nipping or biting is to press its head down with your hand for about five seconds. If it still tries to nip, do it a few more times. This demonstrates your dominance to the rabbit—if it feels you are the boss, it is less likely to nip you.

Lastly, you can use a squirt gun or spray bottle to spray your rabbit with water when it exhibits this unwelcome behavior. It's important to be prepared to react immediately with a quick spritz toward the face or head. If you don't, the rabbit won't connect the unpleasant consequence with its behavior of biting.

Pressing firmly on a rabbit's head sends the message that you won't tolerate nipping or other assertive behavior.

Biting Out of Fear or Stress

Your rabbit could be biting because it is afraid or stressed. If your rabbit is trying to get away from you and you forcefully grab it, you may get bitten. It's important to set boundaries with your rabbit while also respecting its personal space, especially if it is nervous around people. If you do get bitten, try to remain calm. A bite is alarming and can hurt, but you can make the situation worse by jumping away or yelling.

If fear was the original cause of the bite, try not to spook your bunny further. Give both the bunny and yourself a little time apart to settle down. Going forward, spending time with your rabbit and acclimating it to human interaction should decrease your chances of being bitten.

Sometimes what seems to be fear is actually pain. If a rabbit is not feeling well or is in severe pain, biting is a normal way for it to defend itself or notify you that something is wrong. Pay attention to the rabbit's behavior after a bite occurs and assess whether it is acting normal or showing signs of illness, such as curling into a ball, grinding its teeth, or moving slowly.

Another reason people get bitten is simply because they put their fingers in front of a rabbit's mouth. Rabbits' eyes are placed on the sides of their head, which means they have a small blind spot directly in front of their faces and under their chins. When feeding a rabbit by hand, don't put your hand right in front of its mouth, as it may mistake your fingers for a treat. Instead, hold the food somewhat away and to the side, where it can see better. It's important to teach this to children so that they don't get hurt and your rabbit doesn't learn a bad habit.

If you have trouble figuring out why your rabbit bites, join a rabbit group online and ask if there is anyone local who specializes in rabbit training specifically for biting issues. Sometimes it's best to have a professional assess your situation in person; they can help you understand why the behavior persists.

Offer food from a rabbit's side, where it can see what you're doing, rather than directly in front of its face.

Adding Another Bunny

I am often asked if it's better to adopt two bunnies or if having one is okay. There is no right answer to this question. Rabbits aren't herd animals and don't require other rabbit companions. They are independent and can be perfectly happy with loving humans. Most, however, do very well and thrive with a friend.

Another common question is whether it's better to have two of the same sex or of opposite sexes. I can't stress enough that this does not play a huge role in whether rabbits will bond. You may read online that opposite sexes make better pairs, but it really comes down to personality. Many people believe that females are sassier and males are sweeter, but I say throw that stereotype out! I have had dozens of unaltered females who were extremely sweet and snuggly. I've also had some boys who have been moody and others who have been sweet. Of course, unless you want even more rabbits, at least one member of an opposite-sex pair must be spayed or neutered.

The main thing to understand is that in every pair there is usually a dominant member. It can be a little rough to watch them establish that relationship, but it's best not to interfere unless they are full-out attacking each other. It's normal for them to nip each other, and they may run in circles as they do this. If there is squealing and one of them runs away,

With a barrier between them, two strange rabbits can sniff and get acquainted without feeling threatened.

this is better than if they begin to fight. Mounting is also normal. One may even start mounting the face of the other. Avoid separating the rabbits if you can, because they can only establish a hierarchy, and thus a stable relationship, when they are together. But if they seem to be in serious danger—attacking each other aggressively enough to pull off hair or draw blood—do separate them, wearing heavy gloves so you don't accidentally get bitten.

Two young rabbits adopted together are likely to bond fairly easily.

If you must separate your rabbits, it's best to keep them apart until they are fixed and then gradually bring them back together after they recover. They may take to each other again right away—but you will just have to see how it goes. Don't leave them unsupervised before you are confident they are friends again. If they are still fighting, put them in different pens in the same room so they can see and smell each other. Gradually move their pens closer and closer until they can share a pen wall without reacting aggressively. You can also swap their litter boxes and toys back and forth between pens, moving their scent around to help them get used to one another's presence. The final step is to remove the shared wall or put them in one pen to see how they react to each other. Some will settle down within days while others may take weeks or months. You will have to experiment and see what works best for your buns.

The Younger, the Better

Adopting two babies together is the easiest way to bond them. It doesn't matter if they are from the same litter or even the same breeder. As long as they are both younger than 12 weeks old, they should get along and create a relationship. Some bunnies can start to get hormonal as early as eight weeks, and dominant behavior could arise. They aren't likely to fight when they're that young, but it's better to start creating a bond before their hormone levels go up. As they mature together, one of them will establish a more dominant role. You may have occasional issues during the hormonal stage if one doesn't want to be submissive. It's best to get them fixed as soon as possible to reduce their hormone levels.

If you choose to add another baby bun to the mix soon after adopting your first baby, chances are still good that they will bond easily and quickly. Every rabbit is different, so take precautions when you bring the new baby home. Slowly introduce the two bunnies and make sure your older bun doesn't aggressively attack the new one.

What If My Rabbit Is Older?

If you raise a baby bun for a year or more without other rabbits, it may be a little tricky for it to bond to a new companion. The longer your rabbit is alone, the more territorial and independent it can become, meaning it could reject a new rabbit who enters its territory. You'll have to take the process day by day, carefully watching the way your older bun responds to a new rabbit.

I am often asked to match a younger bunny to the personality of an older one, but no breeder can tell you which of their babies will best pair with a rabbit who has been on its own for some time. However, don't be afraid of the idea of bringing a new baby home to your older rabbit! It often works out just fine. Babies will usually bond easily, so you aren't dealing with two set-in-stone personalities.

If you want to adopt a second older rabbit, it could take a little more time for your two bunnies to bond. An older rabbit is already set in its ways, and if it came from a home where it was the only bun, it may be very territorial as well. There is a much higher chance of fighting breaking out between two rabbits that are older; however, many owners have successfully bonded another older bun to their current one.

Introducing the Newcomer

When you adopt a second rabbit, I highly recommend taking your first rabbit with you to pick up the new one. Use a carrier large enough to contain both of them and put the new rabbit in the same carrier. Once they're in the carrier together, take them to the car right away so they don't have time to start scrapping with each other. Most rabbits do not like car rides, and having a friend to share the experience can help create a bond. The stress of the ride encourages them to huddle up side by side for protection, rather than fighting. This is called stress bonding, and fights in this situation are very rare. If the car ride went well, set the carrier down in a neutral space in the house and open it.

Let the bunnies come out on their own, together or separately. If they aren't that interested in each other, don't worry. This is actually a good sign. Over time, as they start to get more comfortable with each other, they may practice mutual grooming or chase each other playfully.

If you can't introduce your bunnies with a car ride together, introduce them in a neutral area where neither rabbit has been. A bathtub or an outside pen are both good options. Place both rabbits in the neutral space and watch how they interact. If they start aggressively fighting, separate them and take a slower approach. But mounting, nipping, and even pulling out bits of fur isn't grounds for separating them; they are just trying to figure out who is dominant. This behavior can last several minutes to several months. Do not separate them during this period: They need the time together to understand who is who in the relationship. Only separate them if one is dangerously attacking the other.

Once your rabbits establish their hierarchy, the mounting and circling behavior will lessen or stop completely. But be aware that some pairs will never learn to get along and may need to always be housed separately.

When bringing a second bun home, it's best to be prepared with a second pen set up in case bonding isn't going well. Place the pens in the same room so the rabbits can see each other. Side-by-side setups are even better, so they can be near each other without touching. Make sure any shared wall between pens isn't made of wire mesh or any type of screen, as they can bite each other through the holes and may chew through the material. A solid wooden partition is fine, but a clear partition made of an acrylic sheet is ideal, allowing the rabbits to see each other without making contact. While the rabbits are separated, switch their litter boxes back and forth every few days so each can get used to the other's scent

Mounting, nipping, and minor scuffles are a normal part of establishing a relationship between two rabbits.

in its own space. Leave some soiled litter in each box so the smell of each rabbit is strong.

Swap their toys back and forth, too. Rabbits are notorious for chinning things to scent-mark them, especially in their pens, so exchanging toys is another way for two rabbits to become familiar.

Once the two rabbits aren't charging at each other, or if they begin lying next to each other against the shared partition, you can swap the solid partition out for one with a grid that has openings about 2 inches square. See how they interact with this increased exposure to each other. The rabbits will be able to stick their noses through the grid and communicate. They may start licking and grooming each other's face or nose, which is a great sign (and very cute to watch!).

If they try to attack or bite aggressively through the grid, put the clear partition back. Most rabbits will form a relationship within a week or so, but you may have to keep them separated for several months as you work through the process. It is possible that they may never bond and will need to be separated indefinitely, but most rabbits will eventually be able to coexist. Every rabbit has its own personality, which dictates how safe it feels in its home and how territorial it is. We humans must be patient and let them find their own way.

Traveling Companions

If one of your rabbits has to see the veterinarian, always bring both rabbits with you, even if the other one doesn't need to be seen. Riding in the car together reinforces their bond, but more important, they will both pick up the strange smells from the animal clinic. If you leave one bunny at home, it might later attack the other one because it smells so different.

If one of your rabbits needs to stay at the clinic overnight, bring the other rabbit with you when you pick up the first one. Even though the rabbit who was visiting the clinic may smell strange, the car ride should help keep their bond.

Having More Than Two Rabbits

Most rabbits are capable of bonding with another rabbit, but the more you add to the mix, the trickier it can be. There are, of course, exceptions. Several times I've adopted out three baby bunnies together with great success. It is much easier to have multiple baby bunnies go to the same home, where they can establish their territory together before their hormone levels rise. This doesn't mean fights won't ever break out; you may still have a period of time where they need to be separated until they are spayed or neutered. However, the younger they are, the easier the bonding process should be.

There have been instances where someone adopts two bonded rabbits and later wants to add another bun because they're enjoying them so much. This isn't impossible, but a lot can happen when you change the dynamic of a pair of bonded rabbits. The original pair could break their bond or even attack each other because of the disturbance. There is also a chance that one of the bonded pair could start bonding to the newcomer, leaving the third rabbit frustrated to the point that fights ensue. This isn't

Rabbits don't mind being solitary but they are also capable of living in pairs, trios, or even larger groups.

always the case, but it's good to know that it can happen and to be very watchful of your bunnies' behaviors during the first few weeks.

If you want a third rabbit, you have the best chance of success with an eight-week-old baby. This means you aren't trying to bond three solid personalities but rather introducing a youngster who is not yet hormonal or territorial. If you choose to adopt an older rabbit, look for one who is already fixed or get it fixed as soon as you can. Bonding spayed and neutered rabbits is usually more successful because of their reduced hormone levels.

I don't recommend adopting a third rabbit who is three to five months old, as this is when rabbits typically are the most hormonal. If fights break out, it could be hard for the rabbits' relationships to recover. If they mark each other as enemies, it can be difficult to get them to bond in the future. However, car rides in a shared carrier could help. Take them on multiple car rides throughout the week; when you get home, place the carrier in neutral territory, let the rabbits come out on their own terms, and see how they behave toward each other.

Introducing a Third Bunny

The best way to introduce a third rabbit to a bonded pair is with a separate pen that shares a side with your other rabbits' enclosure (see Adding Another Bunny, page 80). Watch over the next few days to see if either of the bonded rabbits charges at the wall, or if there are any disturbances between the pair. If all seems calm, take the rabbits to a neutral space in the home where none of them have been before—or even set up a pen outside—and put all three together with plenty of toys. Watch closely for aggressive fighting. Again, mounting and circling each other is normal; don't separate them unless aggressive fighting breaks out.

As a last resort, you can try to board all three rabbits at a facility for about a week; this puts them all in a neutral space. While they are being boarded, clean out their pens at home and sanitize everything to remove existing smells. It's even better if you can reconfigure their area or move it to a different part of the home. Coming home to an unfamiliar space may prompt the rabbits to reset their territorial and aggressive behavior. You might also find a specialist who works with people to help their rabbits bond.

Introducing Rabbits to Other Pets

If you already have pets, particularly dogs or cats, ask if the rabbit you are buying or adopting has been exposed to other animals. Even if a rabbit hasn't been around other types of animals, if it is well socialized and your other pet does not have a high prey drive, there is a good chance they can coexist peacefully, depending on both animals' personalities. Be honest in assessing your dog's personality. A dog with a high prey drive may never be able to live comfortably with a rabbit.

If You Already Have a Dog

I recommend looking for a bun who has been raised around dogs and isn't intimidated by them. Rabbits who have been raised around dogs tend to be bolder. If you can't find a reputable breeder who socializes their rabbits with dogs, find one who spends time socializing their buns well with people. If your dog has never interacted with a rabbit before, introduce the animals carefully to see how they react to each other.

If you adopt a pretty social bun, give it a few days to adjust to its new environment before slowly introducing it to your dog. A very timid bunny will need a lot more time to adjust to a new home before meeting a dog. Trust is everything in a relationship with a rabbit. If your rabbit feels safe around you, it will interact more with you. The same goes for your dog. One thing to remember is never, and I mean *never*, leave your rabbit alone with your dog. No matter how bonded they are, animals can be unpredictable, so always supervise them!

If your dog is calm around small animals, you can introduce them by sitting on your couch with your rabbit on your lap. Let your dog approach and sniff or lick your rabbit gently; watch how your rabbit responds. If it starts to freak out, scratch, and try to get away, end that session and try again another day. The goal is to make sure your rabbit feels safe interacting with your dog. If the dog barks or gets too excited, give them both more time to get used to the sound and scent of another animal in the house.

A slower but safer method is to set up a pen in a neutral area where your rabbit can hop around without interference. Place it in a corner so your dog can't run around the perimeter, and provide a box or bed that your rabbit can hide in if it feels nervous. After your bun is used to the pen, let your dog into the same room so it can approach and sniff the rabbit through the enclosure.

Start with short daily sessions and watch to see how the animals are doing together. Some dogs will be a little too excited and will pace back and forth along the pen. This is okay as long as your rabbit is not scared. As the days or weeks go by, if your dog shows less interest or is calmer near the pen, you can start slowly introducing them face-to-face. Although the pen method takes more time, it can be more effective because it allows the pets to get used to each other while maintaining a safe boundary. If your dog doesn't calm down and still tries to chase the rabbit, be prepared to accept that it might not work out the way you want. Don't keep forcing the relationship because it could cause your rabbit a lot of trauma and stress. You may have to keep them separate always.

If You Are Adding a Dog

Rabbits can form bonds with companionable dogs and cats, and even other animals.

If you already have a bun and want to get a dog, it's best to find a dog who has been exposed to other small prey animals. You won't always know this if you are adopting from a shelter or rescue organization, but these groups often test dogs to see how they react to other animals. A dog who has lived with cats before could be a great option. Breed plays a role as well, so avoid breeds that are known to have a high prey drive.

Adopting a puppy could be your best bet because as its personality develops you can train it to share space with your bun. Young puppies sleep about 90 percent of the day, and, believe it or not, this is the best time to socialize them with your rabbit. When your puppy is sleeping soundly, put your rabbit directly next to it. Even asleep, the puppy will become used to the rabbit's scent. When the puppy wakes up and first sees the rabbit, it will be groggy. It will probably be curious and start sniffing and licking, which is a good sign. Your rabbit is less likely to be afraid if your puppy has been lying still, and if the rabbit is calm, the puppy won't be triggered to chase it. Repeat this process a number of times, and your puppy will begin to regard the rabbit as a friend, not food. Remember, though: Even if your puppy has known rabbits all its life, don't ever leave them unsupervised together.

What About Cats and Other Pets?

Cats and rabbits can form friendships if introduced carefully; I know several families with rabbits and cats who live happily together. Depending on their personality, cats, like dogs, will react differently when introduced to a rabbit. Many cats may want nothing to do with a rabbit, some may be afraid, and others might be curious and eager to meet the newcomer. However, don't forget that cats are predators and some of them might attack a rabbit, especially a small one. Always be present while a predator and prey animal are in a room together. Every animal is different, and only time will tell if they can coexist.

Other animals can bond to rabbits as well. We once adopted a bunny to a family with a house pig. The two of them napped together on a dog bed and got along great!

CHAPTER 4

Feeding Your Rabbit

RABBITS HAVE VERY SENSITIVE DIGESTIVE SYSTEMS, and they must eat a high-fiber diet to keep everything working smoothly. Contrary to the popular notion of salads as "rabbit food," lettuce and raw vegetables should be only a small part of a rabbit's diet. Like horses, rabbits should mostly eat hay.

RABBITS ARE HINDGUT DIGESTERS, which means that in addition to their stomach and intestines, their digestive system includes a cecum, a pouch where hard-to-digest grasses and other cellulose are processed by specific bacteria. The contents of the cecum are passed as cecotropes, soft droppings that the animal eats, usually soon after passing them. This two-part digestion process, known as cecotrophy, allows the rabbit to absorb important micronutrients and maintain a healthy balance of microflora in the gut. Rabbits also pass harder, round droppings that they do not eat. Eating cecotropes, though it may seem gross, is critical to your rabbit's health, so don't remove them.

CECOTROPES

NORMAL DROPPINGS

Feed Hay Every Day

Hay is the most important food for rabbits because it is high in fiber and chewing it helps keep their teeth from growing too long. They must have access to unlimited hay at all times. Each day, your rabbit ideally should eat a pile of hay the size of its body. It's always a good idea to check with your veterinarian to see what kind of hay your rabbit should be eating. If your rabbit has any health issues or sensitivities, your veterinarian will be able to help choose the diet that fits best.

You probably know that hay is long grass that is cut and dried before it is gathered into bales. There are different kinds of hay, not all of which are ideal for rabbits. Mixing hays can be a great way to offer your rabbit different textures and flavors. Try combining oat hay with timothy or orchard grass for a balanced blend. You can mix any two hays equally. If you want to mix orchard grass with timothy, for example, offer a 50-50 mix. If you want to add another textured hay such as oat hay with timothy and orchard grass, offer it in three equal portions. The following sections describe some different types of hay.

Hay can be offered loose or in a feeder. Either way, give your rabbit plenty of it.

Alfalfa hay is high in calories, protein, and calcium. It should be fed to baby bunnies to support the rapid growth of the first six months. Alfalfa is too high in calcium for adult rabbits and can cause health issues such as kidney problems. Although you can feed babies only alfalfa, doing so may make them picky about trying new kinds of hay as adults. I prefer to feed babies an alfalfa and timothy mix, which makes it easier to transition them to timothy hay after six months. The last thing you want is for your rabbit to refuse to eat different hay because it only wants alfalfa.

ALFALFA HAY

Timothy hay is the best choice of hay for adult rabbits. It has just the right balance of protein, calcium, and fiber. A high-fiber hay such as timothy should make up the bulk of your adult rabbit's diet, but you can mix in other types of hay to offer a treat or to tempt a picky eater.

TIMOTHY HAY

Meadow hay is a mixture of wild-harvested grasses. It can include herbs, flowers, and even twigs, and is said to be the most environmentally friendly hay out

MEADOW HAY

there. The only downside is that it's not as high in fiber as timothy hay, and a picky rabbit could end up eating only the sweeter pieces of the meadow hay and leaving the higher-fiber stems alone. Always start out offering higher-fiber hays before resorting to meadow hay.

Oat hay is a good source of fiber, but it is not a grass hay, which is what your rabbit should primarily consume. Rabbits typically enjoy the extra-crunchy texture of oat hay stems and immature seedheads. This hay is higher in fat and carbs, so it should not be the only hay for your rabbit. If you decide to broaden your bunny's nutrient intake, it's safe to create a blend of about 20 percent oat hay and 80 percent grass hay.

Orchard grass is a very soft hay with a sweet taste that most rabbits enjoy. If you notice that your rabbit is not eating much timothy hay, try mixing in some orchard grass, whose sweeter taste and finer stems may encourage it to eat more.

First- and Second-Cut Hay

Producing quality hay takes skill, the right soil, and proper weather conditions that ensure proper drying so it doesn't end up scorched or moldy. Hay can be harvested more than once per season and is classified as either "first cut" or "second cut."

Most farmers harvest their first cut in early summer. First-cut hay has thicker stems and fewer leaves, giving it the most fiber and helping keep your rabbit's teeth ground down. First cut also has less fat, so it's the best choice for an overweight rabbit. The only negative thing about first cut is that some rabbits prefer to chew softer hay. If your rabbit isn't a big fan of first-cut timothy, you can buy a different cut to see whether your bun likes it more.

Second-cut hay is harvested within two months of the first cutting. This is also a great option because it has slightly more protein than first cut and the stems are a little softer, making it easier to chew. Second-cut hay also has more leafy greens that rabbits love.

Sometimes there is a third cut, which is like pure gold to most rabbits but is often hard to find. Most farmers don't risk a third cutting in the fall because wet weather is more likely to damage the crop. This is the softest cut of hay—leafy green with minimal stems—and is higher in fat and protein. It's ideal for bunnies who are very picky or need to gain weight.

Finding Good-Quality Hay

Not all hay is of the same quality, and you may have to look around to find a high-quality hay. It's worth the search because, even if it costs a bit more, good hay is better for your furry friends and can actually save you money because there will be less waste. You can buy bagged or boxed hay at pet stores and feed stores or order it online. If your rabbit isn't eating its hay, try different brands or different cuttings. Some companies offer a sampler pack of different types of hay, which can be a good way to figure out what your rabbit likes.

If you live in a rural area, ask local farmers or horse barns if you can buy a bale or two at a time, if you have room to store it. Hay must stay dry, so keep it in a garage or a shed, off the ground. Also, to preserve the nutrients, make sure to keep it out of direct sunlight. You can store a whole bale in a large plastic bin with a lid; make sure to drill holes for air circulation. Baled hay is made of flakes (layers) that come apart, so you can also get several smaller bins that will hold three or four flakes and stack the bins to save space.

Hay cubes, which are small (about an inch or so), tightly compressed blocks, are a great alternative to bales, especially if people in your household are allergic to hay, which can be dusty. If you have a picky bun that has rejected different types of hay, the cubes might entice them to eat more. Start by giving five or six cubes per day and adjust accordingly. If all the cubes are gone at the end of the day, add more the next day; if there are some left over, reduce the number of cubes. You can also offer one or two cubes along with unlimited loose hay as a treat and a good way to wear down teeth.

BALE

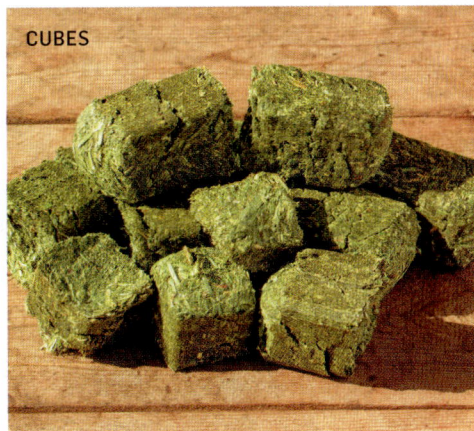

CUBES

Offer Pelleted Food for a Balanced Diet

In addition to unlimited loose hay, I recommend a high-quality pelleted food for your rabbit. There are a lot of brands available with many different ingredients. So what makes a pelleted food high quality? First, the main ingredient must be hay. Pellets for young rabbits are usually alfalfa-based for higher protein, calcium, and calories. An adult rabbit should have a timothy-based pellet unless your veterinarian recommends otherwise (some elderly rabbits get alfalfa to help with weight gain or sensitivities). High-quality pellets also contain essential elements such as fiber, proteins, vitamins, and minerals.

Rabbits younger than six months old should have access to unlimited pelleted food unless they have soft droppings. If they have an excessive amount of soft poop, limit them to about a quarter cup of pellets per day to encourage them to eat more hay. For rabbits older than six months, follow the feeding instructions on the pellet bag.

Some owners choose to eliminate pelleted food completely and feed their rabbits simply vegetables and hay. While you most certainly can feed your rabbit this way, most owners find it more convenient and cost-effective to use pellets. Pelleted food is formulated to meet a rabbit's daily needs and is easy to measure. I've used pellets for more than 25 years and have never experienced any issues. In fact, most rabbits really enjoy the texture and flavor of pellets, and it's an easy way to know your rabbit is getting all the nutrients it needs. If you plan on going pelletless and feeding your rabbit only fresh greens and hay, ask your veterinarian which vegetables and hay mixtures will ensure it gets all the necessary nutrients.

While most pellets on the market list hay as their first ingredient, it's also good to know what should *not* be in your rabbit's pelleted food. The pellets should be greenish and compact,

Choose high-quality hay-based pellets without the added treats shown on the left.

with no nuts, seeds, fruits, or colorful pieces. These unnecessary ingredients are usually high in sugar and carbohydrates, which can cause health issues and obesity. Most rabbits offered this mixed feed will pick out the high-sugar items and leave the healthiest parts behind. They will then wait for you to refill the bowl so they can pick through it again for the yummy treats. Stay away from these feeds and stick with a hay-based pellet that contains the proper balance of nutrition and leaves out sugary extras.

Be sure to measure pelleted food so your rabbit doesn't overeat.

How Much and How Often to Feed

Rabbits should consume about 85 percent of their diet as hay, supplemented with pellets. A good pellet ration is ¼ to ½ cup per six pounds of your rabbit's weight per day. Follow the guidelines on the bag unless your veterinarian recommends otherwise. Don't worry about feeding too much or too little unless you notice a change in weight. If you notice weight loss, first confirm that your rabbit is actually eating its hay. If it's consuming sufficient hay but is losing weight, add a few extra tablespoons of pellets to its daily diet and see if this helps. If your rabbit is overweight, cut back on the pellets (eating mostly hay shouldn't make your rabbit gain weight, so if this happens it's likely from too many pellets). If you notice sudden or persistent weight issues, ask your veterinarian to help you figure out what's going on.

Unlimited hay should be available at all times. Pellets can be given all at once or divided into two or three daily servings. The main thing is to find a system that works with your schedule and stick to it. Don't worry if your rabbit eats all its pellets in the morning and has none left later in the day. If it's hungry, it will eat hay, which should be the bulk of its diet anyway.

Feeding Fresh Greens

Can your rabbit live a long, healthy life without eating fresh greens? Absolutely. There's a lot of information online that insists rabbits must have fresh greens daily, but I disagree. They should consume hay primarily, with pelleted food added for balanced nutrition. Over the years I've met several veterinarians in exotic-pet practices who have differing opinions on this subject. Some claim it's essential for rabbits to eat greens, while several others claim that a high-quality hay and pellet diet is sufficient and appropriate.

I have tried different feeding methods for my rabbits over the years and have noticed no difference when I fed pellets and hay versus adding a variety of greens to a hay-based diet. Whether you feed your rabbit fresh greens is your choice, so don't let anyone shame you if you decide not to. Fresh food holds more moisture than pellets, which helps keep your rabbit hydrated; however, your rabbit will drink plenty of water if you decide not to offer fresh veggies.

Whatever you do, always introduce new foods one at a time with at least a week between them. That way, if your rabbit has a bad reaction to something, you'll know what the problem food is and can avoid it in the future. If you're offering new greens every couple of days, it can be hard to tell which one caused an issue. When feeding fresh veggies to a youngster less than six months old, start small, as its digestive system is still sensitive. Eating too many vegetables can cause gut issues, which, if left untreated, could be fatal.

I tell my customers to use their pinky as a size guide when feeding vegetables to baby bunnies—that is, only give babies a total daily amount of vegetables equal to the size of your smallest finger. Once they are five months old you can increase this slightly, and at six months they can

have a few cups of fresh greens per day, depending on their size. If you want to wait until they are six months old to introduce greens, that is fine as well.

You can feed your rabbit fresh greens in the morning only or split them between two or three servings throughout the day. One to two cups of greens daily per four pounds of your rabbit's body weight is standard. Also note that it's always a good idea to rinse off greens and other vegetables prior to giving them to your rabbits. Watch your rabbit's stool to make sure it remains hard round pellets. Any signs of diarrhea should be taken seriously; if you see this, cut out all high-sugar vegetables immediately. Increasing fiber will harden your rabbit's stool again.

If you want to give your adult rabbits kitchen scraps such as carrot tops or lettuce cores, that's fine. Just be aware of which vegetables are not safe and remove uneaten food from the cage every day.

Safe greens include lettuces (except for iceberg, which can cause bloating from its high water content and low fiber), arugula, collard greens, bok choy, mustard greens, carrot tops, and spinach, as well as cilantro, basil, parsley, mint, dill, thyme, oregano, sage, lavender, fennel, and rosemary. Other safe vegetables include celery, cucumber, zucchini, and radishes (with the tops).

Rabbits enjoy greens and other fresh foods, but don't give them too much at a time.

Some vegetables should be treated carefully because they have more sugar content and may not be as high in fiber. For instance, carrots are high in sugar and should only be given sparingly as a treat (the greens are fine). In addition to carrots, you can offer your bunny small pieces of Brussels sprouts, parsnip, sweet peppers, kale, beetroot, squash, pumpkin (also used to unblock the digestive system), cauliflower, and broccoli. Don't give your adult rabbit more than a couple of one-inch pieces of any of these vegetables each day.

Foods to Avoid

There are many foods you should steer clear of to ensure your rabbit has the correct nutrition. Many treats labeled rabbit-safe—even advertised for rabbits specifically—may not actually be good for your bunny. Processed foods, preservatives, and sugar-heavy treats should not be part of your rabbit's diet. Dairy is not safe for rabbits, so stay away from cheeses, yogurts, and any other dairy-based products, which can cause stomach issues—this includes yogurt drops, often marketed in pet stores as "safe" for rabbits.

Nuts and seeds are not toxic to rabbits, but they are very high in fat and can cause digestion issues if eaten too often. (I would recommend leaving them out of your rabbit's diet completely, but if your rabbit accidentally gets into some nuts or seeds, it will most likely be just fine.) Cabbage, broccoli, or cauliflower can be given sparingly, but too much can cause gas and bloating, which can seriously affect your bunny's digestive tract, so don't make these foods a part of their daily or even weekly diet. Or just opt out if you want to avoid any possible issues!

In addition, do not feed your rabbit the following:

- Chocolate or anything containing caffeine
- Bread, pasta, cookies, crackers (anything overprocessed)
- Peanut or other nut butters
- Meat
- Potatoes, including the peels
- Beans or any legumes
- Corn, including the husks
- Avocado
- Rhubarb
- Silverbeet (chard)
- Onions
- Garlic
- Apple seeds (these contain cyanide)

What About Fruit?

Fruit is high in sugar and can be fed to rabbits in small quantities as treats but should not constitute a full meal. The safest daily portion is about the size of a quarter—for example, just three berries or a one-inch chunk of pineapple or other safe fruit. If you notice your rabbit's stool getting soft or sticking to its bottom, cut out the fruit completely and offer only high-fiber foods for a while. Once the stool returns to normal, you can add a small amount of fruit back into its diet.

Your rabbit will likely enjoy most fruits you introduce and beg for more. Don't give in! Fruits are not required for a healthy rabbit diet, so if you have any concerns, you don't have to offer them at all. If your rabbit has continuous gut issues, it most certainly does not need added sugar, and in this case it's best to cut out fruits completely. Remove uneaten fruit promptly, as it can ferment, which isn't good for your rabbit's gut.

Safe fruits include pineapple, cantaloupe, strawberries, blueberries, apple, banana, blackberries, raspberries, papaya (also used to unblock the digestive system), pear, and starfruit.

Small pieces of fruit are fine as a treat, but fruit has too much sugar to make up a large part of a rabbit's diet.

Providing Water

As with all living things, hydration plays a huge role in keeping your rabbit's body functioning properly, affecting circulation, digestion, waste elimination, and temperature regulation. Though fresh greens can help your rabbit stay hydrated, it's imperative that you provide access to fresh water at all times. Rabbits can drink 2 to 10 ounces of water per day. The amount they consume depends on many factors, such as size and activity level. The bigger and more active the rabbit, the more water it will drink.

Warm temperatures increase a rabbit's water consumption as well. If the temperature in your rabbit's environment exceeds 75°F (24°C), it's very important to make sure its water dish stays filled throughout the day. Putting ice in its water bowl can help your rabbit cool off and entice it to drink more. If you worry your rabbit is not hydrated enough, you can give it vegetables that are high in water content such as cucumber, lettuce, or celery.

Bowl or Water Bottle?

Depending on your rabbit's setup, you can use a bowl or a bottle for water. Both are inexpensive, so try different options to find what works best for your rabbit.

If you use a bowl, make sure it's a bottom-heavy ceramic dish that will not tip if your rabbit rests its feet on the edge while drinking. Rabbits can also knock lighter bowls over or throw them around with their mouths, making a mess. Some water bowls have an attachment on the side so you can mount them to something, which eliminates that possibility. Many people claim that rabbits get more water from bowls than water bottles, but this is absolutely not true. A thirsty rabbit will drink from either until it's satisfied. I've never had a problem with water bottles, and in fact I prefer them over bowls for many reasons.

Water bottles keep the water clean from debris or fur, and they can't be knocked over. I feel more comfortable knowing that my rabbits

Whether you choose a bowl or a water bottle, make sure your rabbit always has clean, fresh water available.

You can supplement your rabbit's water intake with treats of celery, cucumbers, and leafy greens.

can't spill their water in the night. Even if a bottle is mounted incorrectly and falls over, there's very little mess. But the main reason I prefer water bottles is that they need a good cleaning only once a month, rather than several times a day (bowls continually collect droppings, fur, or hay pieces). If you decide to use a water bottle, that monthly cleaning is important to avoid mold. I also add 10 drops of apple cider vinegar to each of my rabbits' 32-ounce water bottles on a cycle of three months on and three months off, which helps keep their water bottles cleaner much longer while offering some protection against internal parasites.

CHAPTER 5

Rabbit Housing

RABBITS CAN BE HOUSED INDOORS OR OUTDOORS, depending on the climate. They can thrive in either situation, as long as they have proper housing and receive enough attention. Most pet rabbits live indoors, as people enjoy the wonderful presence they can bring to a home. They are adorable to watch, whether they are running and binkying around or luxuriously sleeping in funny positions, and litter box training is better understood than it used to be.

Some owners still prefer to house their pet rabbits outside for various reasons, including the fear of unwanted smells (which occur mostly when a litter box needs to be cleaned more often). If someone in the house is allergic to the rabbit or the hay, keeping the rabbit outside may be necessary.

Housing for an Indoor Rabbit

Most owners choose to keep their pet rabbits inside. I highly recommend this option because it protects them from the weather and makes it easier to spend time with them, which also means you can better monitor their health. If you're worried about having a rabbit inside because of potential odors, don't be. Rabbits are clean animals that don't smell. If you keep the litter box clean, you won't notice any odors. The only exception to this is unfixed males, who are prone to spraying urine to mark their territory and so should be confined to smaller areas until they can be neutered.

The ideal setup for an indoor rabbit is a comfortable cage, where it can rest, eat, and use the litter box, surrounded by a pen made of linking panels so it can be moved. Having the cage within a pen keeps the pen area cleaner. You can also set up a pen with no cage, just a litter box, food dish, water bowl or bottle, and some toys. If you allow your rabbit to roam freely around your house (or a defined area), it still needs a quiet corner where it can find its litter box, food, and water.

If you have the space, a pen offers your rabbit more room than most cages.

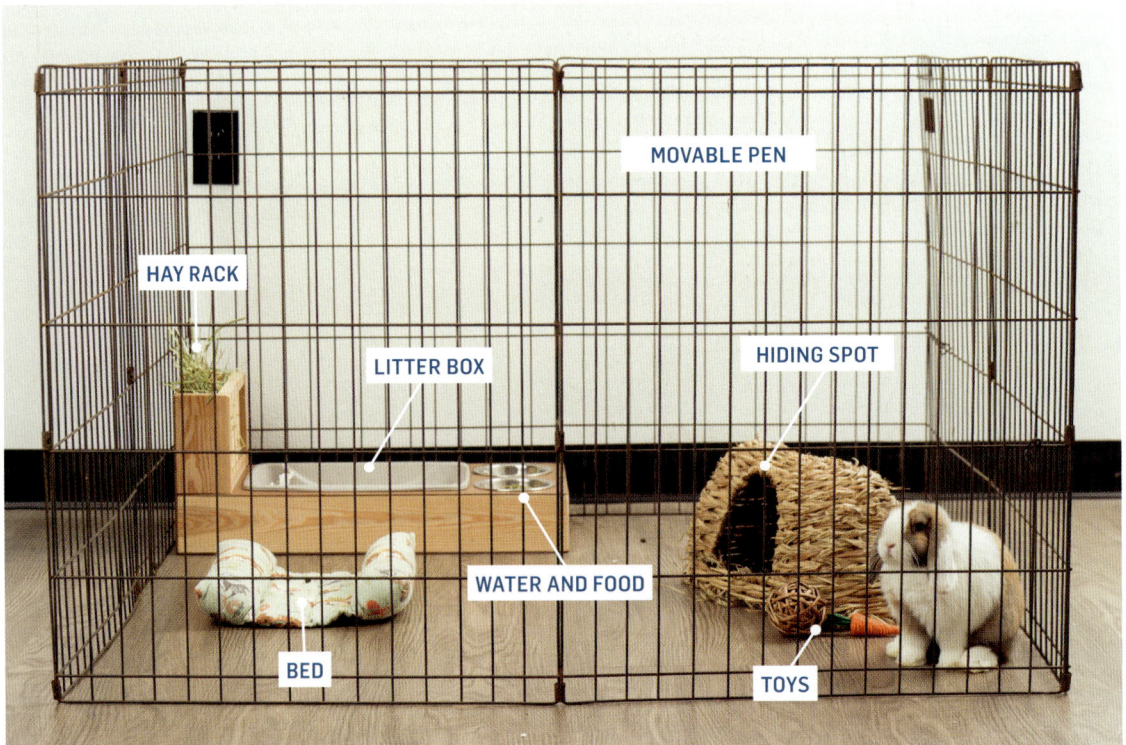

HAY RACK

MOVABLE PEN

LITTER BOX

HIDING SPOT

WATER AND FOOD

BED

TOYS

Cages

A cage is fully enclosed, with a top, a bottom, and a door that fastens. There are many different styles that give rabbits enough space while keeping them and your home safe. However, I advise against most cages marketed for rabbits. They usually measure just 2 × 1½ feet, which isn't enough room for even the smallest breeds to move around comfortably. For smaller breeds, I recommend a cage at least 4 feet wide and 2 feet deep. Larger breeds need cages that are at least 3 × 5 or 3 × 6 feet. I always recommend getting the largest size cage that works in your space, but a smaller cage is fine if you have a pen surrounding it to provide exercise space.

Cages typically come in two parts, with a wire top that snaps onto a plastic bottom. You can easily pop the top off to clean the bottom tray. Compare different cages to figure out what will best suit your home. I like models where the entire top opens, giving plenty of access to reach in and pick up your rabbit. Cages with side doors make it easier for the rabbit to get in and out on its own, but it can be very difficult to lift a rabbit, especially a large one, out of the cage through a small door. You may need to take the whole top of the cage off to pick up your bunny safely.

I don't care for all-wire cages, but they can work fine, as long as the floor is covered with plastic resting mats to protect the rabbit's feet from injury. These mats have holes so that urine drains away. A cage with a plastic floor lining is easier on their feet and much easier to keep clean; you can simply vacuum the cage out or sweep it with a hand broom. If you need to perform a deeper cleaning, you can take it outside and hose it out.

Cages like these are commonly found in pet supply stores, but they aren't the best choice for rabbits. The top one has a ramp, which a rabbit is unlikely to use, and the bottom one is too small for even a little bunny.

Pens

You don't have to keep your rabbit in a cage at all. Pens are a great way to have a cage-free setup. The extra space will let you be creative with the items you put inside your rabbit's area. If you want to make better use of odd-shaped corners in your home, a pen might be a good fit for you: Pens often come as panels that can be connected to make different configurations. Most rabbits can jump higher than 3 feet, so get panels that are at least 3½ feet high. Keep in mind that if you add a play structure, such as a hay bin feeder or castle, your rabbit will be able to jump on top of it, which will make an escape easier. If need be, you can raise the walls with added materials, such as acrylic or wood panels, or cover the pen with a fitted sheet. But paying for a taller pen often makes more sense.

If you are crafty, you can most certainly make your own pen from nontoxic wood, an option that offers endless opportunities for customizing pen shape and size. Pens made with acrylic panels are a little pricier, but rabbits cannot chew through these panels and the material needs minimal cleaning.

Gridded wire pens, usually with 2-inch-square openings, are another popular style. Walls on these pens are usually not very tall, and they may need to be stacked two or three panels high so your rabbit can't jump out. Such pens are inexpensive compared to those with acrylic panels, and you can buy several sets of walls to create a custom pen area. Connecting the gridded panels can be time-consuming, and you'll want to reinforce your build with zip ties in addition to the connection pieces provided. Zip ties are especially helpful when you are stacking walls on top of each other. If you choose to go this route, be especially careful to make the walls too high for your rabbit to jump—if they get their feet caught in the wire grid, they can break their legs.

Flooring Under the Pen

Flooring is an important part of a pen setup. You can place your pen in an area with tile, vinyl, or laminate flooring, which is easy to clean and will discourage digging and chewing. You can simply wipe the flooring down with a damp rag or disinfect it with a vinegar-and-water solution.

If your pen is in an area with hardwood floors or carpeting, you can buy a piece of laminate or vinyl flooring to protect the area from water and urine stains and prevent your rabbit from digging up carpet. Make sure the flooring extends at least 1 foot beyond the pen on all sides. If the vinyl piece is too small, liquids might run over it and damage what's underneath.

Use waterproof flooring under a pen to protect carpets or wood floors.

Outdoor Playtime

If your rabbit can safely play outdoors, it will enjoy hopping around, digging, and foraging for tasty grass and plants such as dandelions and clover. A rabbit cannot be trained to go for a walk like a dog, but some people put harnesses on their rabbits and let them explore at the end of a leash.

The easiest outdoor-play setup is a pen made of movable panels in your yard; make it large enough for your rabbit to move around comfortably. Provide a hiding place or shelter, such as a smaller cardboard box or even a blanket draped over a corner of the pen, especially if your rabbit is timid.

Do not set up a pen where there is any risk that the grass or other plants have been sprayed with any chemicals. Eating contaminated grass could be fatal for your rabbit. Start with short sessions of 10 to 20 minutes, depending on how comfortable your rabbit seems.

Never place the whole pen in direct sunlight, and be aware that even if you place a pen in shade in the morning, it may be completely in the sun as the day goes by. It's safer to put your pens under trees, both for shade and for protection from birds of prey, which have been known to snatch rabbits from backyards. Putting a cover over the pen adds another layer of protection.

The most important thing about taking your rabbit outside is that you should never leave it unsupervised. An attack by a hawk or loose dog or an unexpected escape would be a terrible consequence of running into the house "just for a few minutes."

Litter Boxes and Litter Options

Indoor rabbits require an area to eliminate. Much like cats, rabbits will quickly learn to use a litter box (see Potty Training Your Rabbit, page 113). Buy a rectangular box, not a triangular one. The triangular ones may seem convenient because they easily fit in a corner of a cage or pen, but they usually aren't big enough and create more mess in the long run.

Buy a box stable enough that it won't tip over if your rabbit puts its feet on the edge. Litter boxes for cats work fine. It should be about twice the size of your rabbit, and bigger is better. The sides should be at least 3 inches high. When rabbits urinate, they lift their back end up a little, so with a low-profile box, the pee will spill and make a mess. You can buy a high-profile cat box and cut down one side for access.

A ferret litter box has three high sides and a shorter side in the front for easy access. This style is useful for smaller rabbits who like to dig in their litter boxes because the high sides prevent most of the litter from flying out. If your rabbit is digging in its litter for fun and making a mess, you can add a grid over the litter to discourage the behavior. You can look up DIY tutorials for how to make your current litter box into a grid litter box. It's inexpensive and you can find most of the materials at a hardware store. There are different grid materials, but I've found that plastic is the most comfortable for rabbits' feet and easiest to clean.

It isn't necessary to use litter in a gridded litter box if you plan to clean it often. That said, a thin layer of litter under the grid will reduce odor and absorb urine, making the box easier to clean. You can also use newspapers or a puppy pee pad on the pullout tray to absorb urine and reduce the chances of spills as you're emptying it.

A gridded litter box with a pull-out pan is easy to clean.

My number-one recommendation is a hay bin feeder/litter box combo that has everything in one place and usually keeps the rabbits' area cleaner. These litter box combos, like the one pictured on page 106, are great because as the rabbit munches on hay, loose bits and pieces are confined within the litter box area underneath it. The rabbit won't step on the hay since it's tucked behind the feeder, so there is less waste.

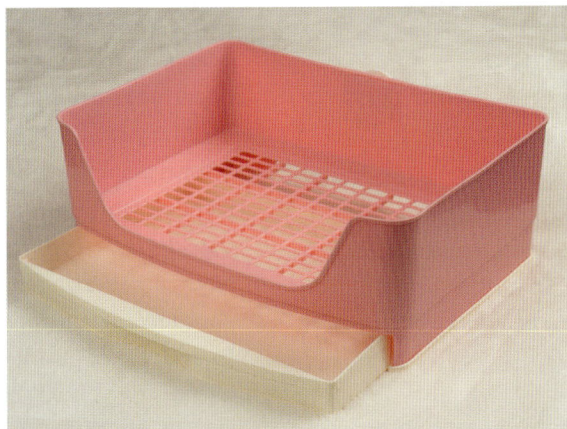

Types of Litter

Litter absorbs urine and reduces odor, which means you don't have to clean out the litter box as frequently. There are several options.

Pine shavings. These are sold in small bags as pet litter and in bales for stall bedding, which can be an economical way to buy them, if you have room to store the larger quantity. Pine shavings are relatively absorbent, but you need to clean wet spots out of the litter box every other day. Do not use cedar shavings, as they can cause respiratory distress in small animals.

Wood pellets. As with shavings, wood pellets are available in large quantities as stall bedding. As they absorb moisture, the pellets expand and break apart. You'll want to empty out the entire box once the poop starts to fill it up, but doing a pee spot-check every other day will save money. You can just scoop out the wet spots and even out the pellets, then add a little more. Wood pellets sold as barbecue fuel work as long as they are not treated with chemicals or made from cedar, juniper, cypress, spruce, or balsam fir.

Paper litter. There are many choices for paper litter, and not all are equal. Look for a type that has clumps rather than shreds of paper. A thick clumping litter absorbs odors better and needs to be changed less frequently. Thinly shredded paper bedding quickly becomes soaked and needs to be changed much more often. Pellet litters made of compressed paper work comparably to the clumping type.

Many owners use a paper shredder to create a supply of litter from waste paper. Shredding newspaper may seem like a cheap solution, but the thin strands become soiled very quickly, meaning you have to change out the box every day. Newspaper does not provide any odor control either. Though shredded newspaper isn't very effective, there are brands of litter made of pelleted newspaper, which absorbs moisture better.

PAPER LITTER

WOOD PELLETS

Warning: Do not use kitty litter or any clay- or sand-based litter for rabbits. These tend to be very dusty and can cause respiratory issues.

Potty Training Your Rabbit

Potty training a rabbit can take anywhere from one day to several months. It depends on your bunny's age and hormone levels and how well it adapts to a new home. Ask the person you adopted your rabbit from for a small baggie of your rabbit's poop; placing this in the new litter box will make it smell familiar and may help your bunny figure things out quicker. It's also a good idea to put some hay in the litter box. Bunnies like to poop where they eat, which is why a combination hay feeder/litter box works well.

When potty training your rabbit, begin by limiting its free-roam space. If you have a bigger pen, make it smaller by putting up a divider or removing some wall segments. Confine your rabbit to this smaller space for the first few days or weeks, depending on how well it is using the litter box. As the rabbit progresses in its training, you can open the space up more. When your rabbit is fully potty trained, you can let it free roam or access its entire pen. If it starts having accidents, restrict the space again until it's using the litter box reliably. Some owners have multiple litter boxes depending on how much space the rabbit can access.

Like cats, rabbits are naturally inclined to use a litter box and can typically be trained within a few days.

Outside the Box

Here are a few tips for those bunnies who may think "outside the box." If your rabbit has an accident, wipe up the urine with a paper towel and place it in the litter box; do the same with stray poops. This reinforces the message that the box is the place where it should do its business. Keep in mind that rabbits are generally not 100 percent perfect with their litter boxes. It is normal for them to leave a few droppings next to the box or to urinate over the edge. Having the litter box on a plastic mat or tile floor makes it easier to clean up these little mistakes.

If you can see your rabbit about to eliminate outside its litter box (it may lift its tail or shimmy down into a seated position), try to pick it up and put it in the box.

If your rabbit is pooping or spraying pee everywhere, this is probably because it is marking its territory. Be patient with a young rabbit who is suddenly not using the litter box reliably. This is most likely due to a rise in hormone levels and will resolve once the rabbit is fixed or grows out of this adolescent phase. Usually, territorial marking stops within a few weeks of a rabbit getting fixed, but it can take up to a couple of months.

Finally, if your bunny insists on going in a particular spot, sometimes it's easier to give in to its preference and move a litter box to that place.

Where to Buy Supplies

The main places to buy rabbit supplies are feed stores, pet stores, or online. Pet and feed stores are similar, but feed stores cater primarily to farmers and may sell supplies like hay in larger quantities. Otherwise, feed stores usually have the bare-minimum essentials. Pet stores focus more on household animals, selling essentials such as feed, small quantities of hay, litter boxes, and litter. Both types of stores may have a small selection of toys, but there usually isn't a lot of variety. If you're looking for unique structures and toys, you'll find the most options online.

Online retailers offer a lot more of everything rabbit-related, often at lower prices, and most people prefer to have everything shipped to their front door. Online subscriptions are a great way to keep yourself stocked with your pets' food and favorite toys, and most allow you to pick how often you'd like your items shipped. If you're on a tight budget and don't mind used items, check online ads for people in your area selling rabbit-related supplies. You may find like-new items for a fraction of the cost. Just make sure to give anything you buy a good scrub-down before using.

Your breeder or shelter should be also able to recommend products and tell you where to find them. Established breeders often link recommended supplies on their websites or may even sell them themselves.

Cleaning the Cage

The number-one priority besides keeping your rabbit fed and watered is keeping their area clean. Rabbits truly are clean animals, but sometimes they can be messy with their hay, or they may dig in their litter box if it isn't gridded. It's important to clean the litter box regularly and sweep up messes, to keep your bunnies from tracking pee and litter box odor around their pen. Not only does this soil their feet, but once their whole pen smells like a litter box, they may start peeing and pooping everywhere. If the box gets too dirty, your rabbit will stop using it.

Another important reason to keep the litter box clean is that rabbits constantly groom themselves with their paws. If their feet are soaked in urine, they'll spread it over their fur, possibly causing an infection, say to their eye, if they happen to scratch themselves. The buildup of ammonia from urine can also create respiratory problems. A clean cage will help you enjoy your bunnies' presence in your home.

It's much easier to spot-clean every day rather than scrub up a big mess once a week. It takes me just half an hour to clean 32 pens, so imagine how much less time it takes to clean a single pen! I use a shop vacuum to suck up stray poops and hay strands; for one or two pens, a dust broom or a hand-held vacuum works fine (if your rabbit isn't used to loud noises, take them to another area until they become accustomed to the sound).

I also wipe down the pens a few times per month with a solution of 1 part vinegar to 4 parts water. You can sprinkle baking soda on the area you'd like to clean and then spray the vinegar-and-water solution over the baking soda, which causes a fizzy reaction that loosens particles and helps sanitize surfaces. I also like to cut a lemon in half and use the exposed part to scrub in circles over the baking soda concoction. This is a safe method for cleaning that will not harm your rabbit in any way. You can find nontoxic, animal-safe cleaning products online or in a pet store, but read the labels carefully and avoid products containing phenols, ammonia, bleach, alcohol, benzalkonium chloride, and synthetic fragrances.

Can My Rabbit Live Outside?

Rabbits do fine living outdoors, but interacting with and caring for them can be harder for the owner.

Rabbits are resilient animals and can thrive outdoors, especially in cooler climates. They are quite sensitive to hot weather, though, so if you live in a region where temperatures reach 80°F (27°C) or more throughout most of the year, I wouldn't recommend having an outdoor rabbit. And although I've seen some great exceptions, living outdoors is typically not an ideal situation for pet rabbits—this is because it's harder to socialize and spend time with them, so they are more likely to be ignored or forgotten. These rabbits often end up in shelters because families lose interest, and taking care of them becomes a more difficult chore. If you do house your rabbit outside, please be invested in its welfare and be a good steward of your pet.

The most comfortable temperature range for rabbits is between 45 and 70°F (7 and 21°C). All that fur protects them in cooler weather, and in that temperature range they can easily regulate their body heat. Still, if your rabbits are housed outside you also need to provide protection from wind, rain, and direct sunlight as well as predators.

Keeping Cool

A rabbit's ears are its natural cooling system. Several blood vessels run close to the surface of the skin, and when a rabbit gets too warm, blood flowing through the ears helps dissipate the heat. But rabbits do not have sweat glands, so to release yet more heat they have to pant. Signs of overheating include excessive panting and wetness around the nose. If you notice these, move your rabbit to a cooler place so it does not get heatstroke.

Rabbits outside also dig to cool down because the ground gets cooler the farther down they go.

Sheds

The best setup for outdoor rabbits is a shed-style home, fully enclosed and safe from predators and the weather. You can add a window for light and ventilation, or even a portable AC unit for warm weather. If AC is not an option, make sure the shed has windows for good ventilation. In summer, keep windows open but screened so that nothing can get in and your rabbit can't escape. You can also keep your rabbit cool in the warmer months by laying a few inexpensive ceramic floor tiles around the area. Tiles tend to stay much cooler than wood or carpet, and when your rabbit gets hot it will lie on the tiles. You can also place frozen water bottles on the shed floor for your rabbits to snuggle up to when they feel overheated. You may have to change out the bottles once or twice per day depending on how hot it gets.

In winter, you can plug in a small space heater designed to either run continuously or cycle on and off. Always be very careful when you have a heater plugged in—make sure it can't tip over or be reached by your rabbit. If you can't use a heater in the colder months, pack hay into a corner so your rabbits can burrow into it to stay warm. You can also layer the floor with blankets for them to snuggle into.

Beyond climate control and comfort, you can also consider other shed variations that will improve your rabbit's well-being. I've seen many cute bunny sheds with holes in the side (like dog doors) leading to an enclosed outdoor area. The more space your rabbit has to run, the happier it will be. To keep your bunnies safe, close their door at night or predator-proof your outdoor pen area. And make sure they can't dig out under the perimeter (see Pet Bunny, Wild DNA, page 70)!

Don't Keep Rabbits on Grass or Dirt

Another thing to keep in mind when housing your rabbit outside is to not have it directly on grass or dirt. The soft texture encourages rabbits to dig, and once they do, they won't stop. They can dig down several feet, or even dig under their pen to escape. It's wise to put wire mesh, such as chicken wire, down on any grass surfaces. Because this wire lies flat against the dirt, it won't put pressure on the rabbits' feet the way a wire hutch floor will, and so will not harm them.

Hutches

Sheds take up space and cost money, so they may not be an option for some families. Hutches are typically much smaller and can be stacked two or three floors high. However, I don't care for hutches; unlike a hamster, a rabbit isn't the kind of animal who prefers small tunnels. Rabbits like to run and jump in the air, twisting their bodies and kicking their feet. A hutch can have three levels and look very nice, but if all your rabbit is able to do is run up and down ramps, it isn't able to physically express itself as a rabbit.

Your rabbit's area should be no less than 10 square feet for small breeds or up to 16 square feet for bigger breeds. Other resources recommend much smaller spaces but, in my opinion, that's really not the life for a rabbit, especially since outdoor rabbits may not get the free-roaming time that many indoor rabbits have. Some families have outdoor hutches with pens attached in the front; adding this free-range space is much better than keeping your rabbit confined in a hutch most of the day.

Hutches are my least favorite rabbit housing option because, in addition to being too small, most are made from wood, which rabbits will eventually chew through, and they are more exposed to the weather. Imagine a cold winter night when a storm is brewing—you're cozy inside reading a book, but your rabbit is outside burrowing in straw in its hutch, trying to stay warm through the night! Not really an ideal lifestyle for a pet rabbit. If you do keep your rabbit in a hutch, make sure to have some kind of hidey-hole or sheltered area inside so it can retreat from the wind and cold.

Any wire-bottomed hutch must include some resting spots of other material so that your rabbit's feet are not constantly on wire. Plastic cage mats work well, and you can line your whole hutch with them, which is much easier on the feet. These plastic mats have holes for urine to fall through. You could also have a mix of wood and wire floors in your hutch. Despite all these ways to make it more bearable, I strongly suggest that you avoid keeping your rabbit in a hutch.

In addition to being sized correctly for the rabbit, outdoor housing must be sturdy, weatherproof, and protected from predators.

CHAPTER 6

Rabbit Healthcare

RABBITS ARE GENERALLY QUITE HEALTHY and have few vaccination requirements. Most take care of grooming themselves, but rabbits with long fur need regular brushing, and all rabbits need their nails trimmed. You should be aware of possible ailments and know the signs that your rabbit might be feeling ill.

How Much Grooming Is Required?

A short-haired rabbit's fur requires little attention, but long-haired breeds, including Angoras, Lionheads, Jersey Woolys, and Fuzzy Lops, require daily grooming. It's important to groom these breeds daily; otherwise, they can consume too much of their hair while grooming themselves. This can lead to intestinal blockage, which can run up an expensive veterinary bill or even be lethal.

Grooming a Long-Haired Rabbit

You need to blow or brush your long-haired bunny's fur daily to avoid mats, which are hard clumps of twisted and tangled fur that usually have to be cut off.

Blowing is a very common technique for long-haired rabbits, using an inexpensive tool designed for grooming animals. You can use a regular hair dryer set on low, but it won't be as efficient and the noise might alarm a skittish rabbit. To blow your rabbit's fur, place the bunny on a stable surface. Blow the fur back and forth in short circular motions, starting from the top of the head. The air naturally separates the hairs and prevents tangling. Continue the process over your rabbit's entire body until there is no clumping. Blowing is an easy daily way to maintain your rabbit's fur, but if you skip too many days, you may find the hair so matted with hay and debris that you need to shear its whole body.

You can also trim your rabbit's long hair monthly to keep it short and less of a hassle to care for. Use scissors with curved or blunted ends to cut off any matted fur, as blowing won't untangle mats that have already formed. Avoid cutting the whiskers, but be assured that they, like the rest of the hair, will grow back if you make a mistake. Some rabbits' hair grows quicker than others', depending on genetics, diet, age, season, and overall health. Typically, a long-haired rabbit's fur is fully grown out again in three to five months.

Though these rabbits are pleasing to look at, ask yourself if you're up for daily grooming sessions. You can always keep their hair short, but stay on top of it so you don't run into issues. I recommend watching some videos to learn the proper techniques.

GROOMING BRUSHES

NAIL CLIPPERS

GROOMING GLOVE

Grooming a Short-Haired Rabbit

If you prefer a lower-maintenance rabbit, any shorter-haired breed of rabbit will be a better match for you—think about Holland Lops, Dwarf Hotots, Mini Lops, Netherland Dwarfs, or one of the giant breeds. All require less grooming than the long-haired breeds. Short-haired rabbits still shed on a daily basis, some more than others. Breeds with short rex fur shed the least. All rabbits molt twice per year, shedding most of their fur and growing a new coat. In spring, rabbits shed their thick winter coats to prepare for warmer weather, and in fall they lose their summer fur as their winter coats grow back.

If your rabbits are housed in a temperature-controlled space, molting is usually less extreme, but they will still molt. There are ways to accelerate the process so you don't have an excessive shedder for several weeks. You can use a fine-toothed comb, which helps get the undercoat out quicker.

Most short-haired rabbits are capable of grooming themselves and keep their fur in tip-top shape, but it's good to run a comb over them once a month to remove any small mats or stray hairs that are holding on.

Yellow Fur? Yuck!

Sometimes a bunny will get urine stains, usually on its feet. This can happen during potty training, but it might mean they are spending too much time on soiled litter.

To get rid of yellow stains, mix corn-starch with enough water to form a paste and rub it on the stained areas. Rinse the area with water and pat dry with a towel. This should clear almost all of the yellow off the fur. Repeat a few times if necessary.

Do Rabbits Need Baths?

Rabbits groom themselves constantly. Do not give your rabbit a bath unless it's absolutely necessary. If your rabbit gets dirty from playing or digging outside, it will clean itself. You can spot-clean your rabbit with a wet wash-cloth, but you should never fully submerge it in water unless for medical reasons or in an emergency situation—it's usually enough to soak only the area needing attention.

If your rabbit has diarrhea stuck to its bottom, for instance, you can hold it while running warm water over the affected area; this helps soften the poop and eventually washes it off. Never use too-hot water, especially in a rabbit's genital area, because the skin is very sensitive and can burn easily. After rinsing the area, wipe your rabbit off with a towel so it is not dripping. It will dry off naturally pretty quickly. Never put a wet rabbit outside in the cold. Rabbits should be at room temperature until they are completely dry.

Rabbits Need Regular Nail Trims

It's important to trim your rabbit's nails regularly to keep them healthy. Like ours, a rabbit's nails never stop growing. Overgrown nails can cause a lot of discomfort, and over time they can even grow into a rabbit's foot pad, causing infection and mobility issues. Most rabbits need a nail trim every 6 to 12 weeks. Try not to go longer than 3 months without a nail trim to avoid issues down the road. You can use the kind of clippers made for cats and small dogs or buy some made for smaller animals. A pair of human nail clippers also works fine.

Rabbit nails have a quick, which is where a blood vein travels through part of the nail. Never cut over the quick, as doing so can cause bleeding. This happens more often than you think, so keep a small jar of a blood-clotting agent such as styptic powder on hand to immediately stop the bleeding if you make a mistake.

If your rabbit has white nails, the pink quick is easy to see through the nail. Cut the tip of the nail just above the pink. If you are nervous, leaving a little white space is fine, but the closer you get, the less often you'll have to trim. It's more important to make sure the tip of nail isn't curling over than it is to cut your rabbit's nails super-close to the quick. With dark nails, have someone shine a light on the nails so you can see where the quick lies.

Rabbits have four toes on each back foot. Their front paws have four toes plus a dewclaw located on the inner paw, a little higher than their other toes. The dewclaw should also be trimmed.

If you are afraid to try trimming your rabbit's nails, you can pay to have a veterinarian or rabbitry do this for you. It's usually inexpensive.

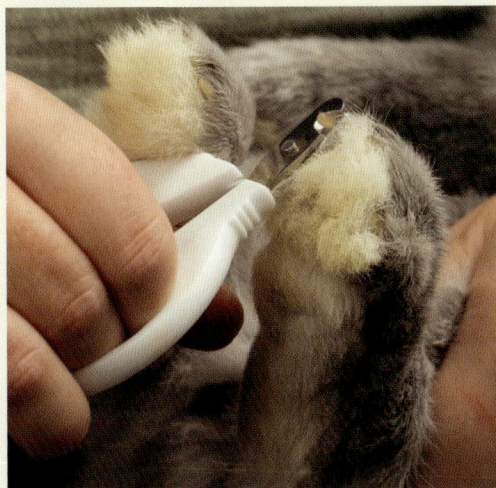

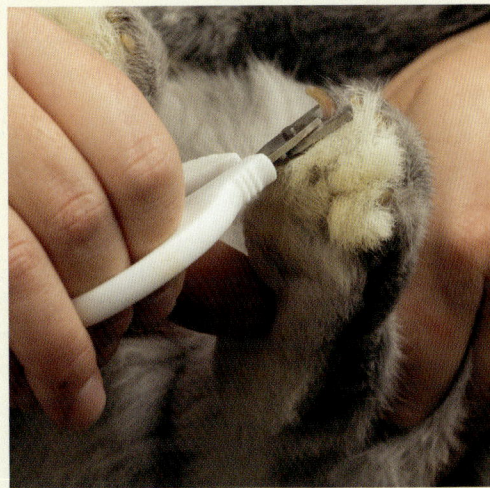

Just clip the tip of the nail, as shown on the left. Clipping farther down the nail risks cutting the quick and drawing blood.

Cleaning Your Rabbit's Ears

Although you can't see all the way down your rabbit's ear canal to check for issues, you can bring your rabbit to the veterinarian for a yearly ear cleaning. You can also purchase a small scope with a camera that connects to your phone and allows you to see down the ear canal. If you see wax buildup, you can break it up with an ear solution for animals. Be sure to follow the directions on the bottle. The process usually involves squeezing a little solution into the ear canal and gently rubbing the ear from the outside to help release the wax. Rabbits will often shake their heads after this, which should also help break apart the buildup. Then you can carefully scoop out the earwax with a specialized tool or a Q-tip. (Just note that sometimes Q-tips can actually push the wax farther down the ear if there is heavy buildup.) It's always good to ask your veterinarian to teach you how to remove earwax correctly, so you don't worsen the problem or damage the ear canal.

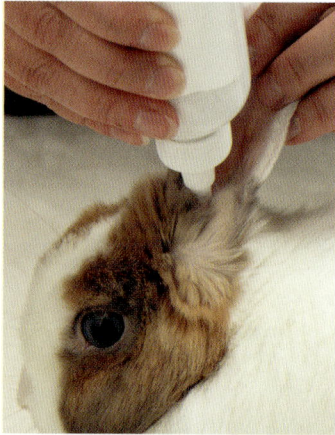

1. Squirt solution into the ear.

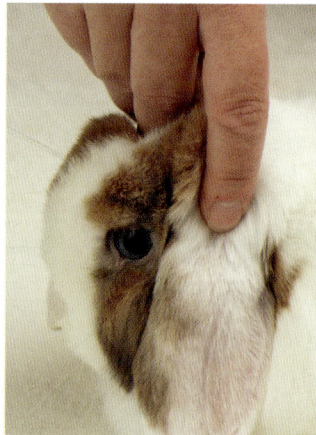

2. Gently massage the solution into the ear canal to loosen debris. The rabbit will shake its head, which helps break up buildup even more.

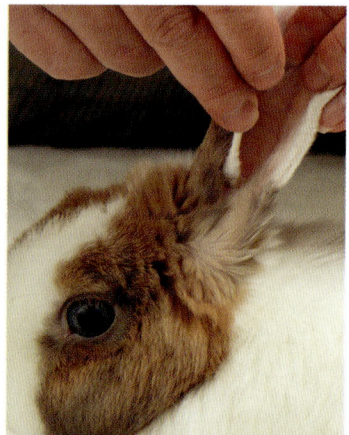

3. If necessary, very gently and slowly use a cotton swab to clean excess wax from the surface of the ear. Don't go too deep.

Healthy Bunny Checklist

Rabbits are generally pretty healthy but as with any pet, it's important to know what is normal and what signals a potential problem. It's better to catch a health issue early than to let it turn into something more serious. As you interact with your bunny, be aware of the following areas and make a note if anything seems out of the ordinary.

Eyes. Eyes should be clear and not leaking discharge or swollen. Some rabbits have a natural light pink lining around the eyelid, which should not be confused with the darker red that could indicate an infection or allergy. It's normal for rabbits to get eye crusties or boogers, and those can be carefully picked away by your clean finger. Excessive gunk around the eyelids could be caused by irritation from something stuck in the eye, or by a mild infection. Usually, regular or medicated eyedrops will help clear the issue quickly.

Ears. The ears should be clean without excessive wax buildup, redness, or swelling. If you have a lop-eared rabbit, regularly lift its ears to make sure the ear crease is not flaky, red, or infected. Mites also like to live near the creases of the ears or back of the neck, so check those two places for any evidence of mites. If they are present, you may need to treat your rabbit with a prescription medication.

Nose. The nose should be clean and clear of any discharge or crustiness. Sometimes lighter-colored rabbits can have a slight green staining on their nose from eating hay. This should not be accompanied by any liquid or discharge. A change in temperature can cause rabbits' noses to water a bit. If your rabbit has been outside or the weather is warm, its nose may be wet to help it cool off. If your rabbit is sneezing excessively, it might have a respiratory infection, which you should ask a veterinarian to diagnose. Some rabbits are allergic to dusty hay and may do better with hay cubes (see page 95) rather than loose hay.

A healthy rabbit looks bright and alert, with clear eyes, a clean nose, and shiny fur.

Respiration. A healthy rabbit takes 30 to 60 breaths per minute, a rate quite a bit faster than a human. A much more rapid rate can indicate that the rabbit is alarmed or simply trying to regulate body heat. You shouldn't hear any wheezing or bubbling sounds.

Weight. A healthy rabbit has a good covering of flesh, meaning you should not be able to see any ribs or backbone through its fur. You should be able to feel those bones when you lightly pass your hand along its body, however. It's better for a rabbit to be slightly overweight than underweight, but if your rabbit is visibly plump, cut back on treats and make sure it's eating mostly hay. If you notice your rabbit losing weight, you might want to try some different foods or take it to the veterinarian.

Fur and skin. A healthy rabbit has soft fur that is clean, shiny, and has no extreme matting, bald patches, or signs of external parasites, such as mites or fleas. Blow through your rabbit's fur so you can see its skin, which should be smooth, not flaky or dry. Flaky or dry skin could be a sign of mites or malnutrition. Mites and fleas are very common and are easily treatable by your veterinarian. Be prepared to treat for mites yearly if they are common in your area.

Feet. Regularly check your rabbit's front paws and back feet for broken nails or injuries. The back feet, which carry most of a rabbit's weight, can develop open wounds; this condition, known as sore hocks, can be very painful.

This rabbit is overdue for a nail trim.

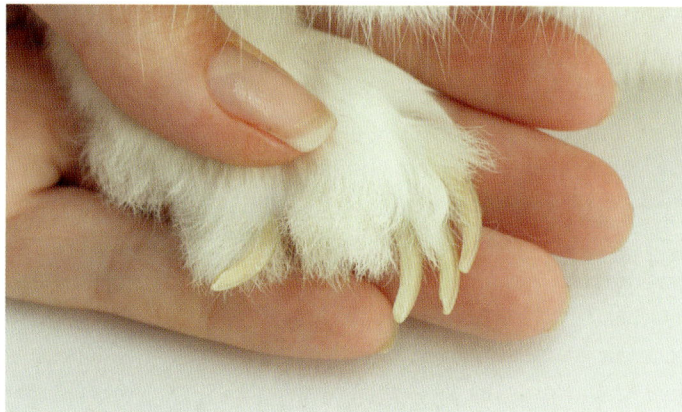

A Sure Sign of Distress

As prey animals, rabbits tend to hide pain or discomfort, so it's good to know the difference between a rabbit who is just hanging out in a comfy loaf position and one who may need some attention. If your rabbit is lying down and taking short breaths with long pauses in between, monitor it closely. When you approach a loafed rabbit and reach out to touch it, it should react alertly, so be concerned if you can gently push on your rabbit and it doesn't respond or even falls over. Other causes for alarm include eyes that stay squinted or closed and poor reflexes when you pick up the rabbit. A weak, listless rabbit needs to see a veterinarian immediately.

Normal Droppings. Rabbits have two types of droppings: solid round ones and cecotropes, which are softer and come out in a little cluster. Cecotropes shouldn't be confused with diarrhea, which is much looser, messier, and smellier. Normal droppings have little odor.

Teeth. A rabbit grinding its teeth is a sign that it may not be feeling well. When rabbits are in discomfort or pain they tend to squint or close their eyes and grind their teeth, which you will be able to hear.

Appetite. Be aware of your rabbit's normal response to food so that you notice when something is not right. For example, if your rabbit is usually excited when you deliver food, you should be concerned if it isn't interested at mealtime. If you notice that hay, pellets, or greens haven't been touched all day, something could be wrong. Consult a veterinarian right away if this behavior persists.

Spaying and Neutering

Spaying or neutering can help make your rabbit a better companion. Your baby bunny may be cute and snuggly when you first bring it home, but between 10 and 16 weeks of age (occasionally as early as 8 weeks), it might start exhibiting hormonal behavior. This can include nipping or biting, aggressively digging and scratching, no longer using the litter box, circling around your feet, and pacing or moving constantly.

If your rabbit shows any hormonal signs, ask your veterinarian at what age they spay or neuter rabbits and make an appointment as soon as you can. Some veterinarians fix rabbits as early as three months old; others wait until six months. Males can usually be fixed sooner than females because their testicles drop at three to four months old and the surgery is less intrusive. A spay is a riskier surgery; the veterinarian must open the rabbit's abdomen to surgically remove its ovaries and uterine horns.

If you are thinking about adding another rabbit to your home, fixing your current rabbit is a must: It will reduce aggressive behaviors and the urge to mate. Usually your rabbit's hormone levels will drop a few days after the procedure, but in rare instances it can take several months.

Veterinarians encourage rabbit owners to get their pets fixed to reduce the risk of reproductive cancers. These cancers are said to be fairly common in rabbits, although in my own experience they have been rare. I've had hundreds of rabbits over the past 25 years and I have never had issues with cancer, nor have I heard about it from anyone who has adopted one of my rabbits. Consult with your veterinarian and use your best judgment.

Cancer aside, getting your rabbit fixed to avoid the possibility of surprise litters is a responsible thing to do. But if you can prevent unintended mating, and you are okay with your rabbit's behaviors as they are, getting your rabbit fixed is not a requirement. The procedure carries risks, especially for females, so if you're able to avoid it, that's ideal. Just be very careful about bringing another rabbit into the household, even if you think they are the same sex! Accidents happen all the time because people aren't educated about how to check sexes. It's easier to check the sex of a fully mature bunny, as males have noticeable testicles.

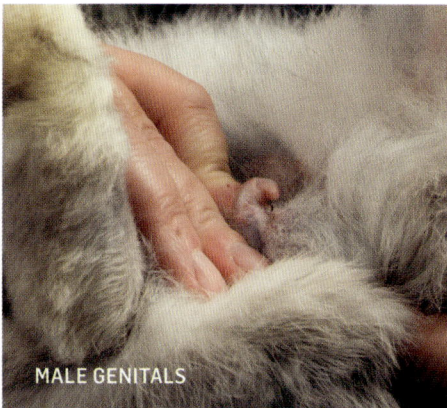

MALE GENITALS

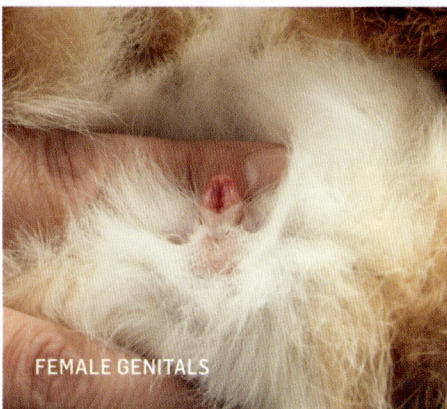

FEMALE GENITALS

To check the sex of your rabbit, use one arm to hold it in your lap, tucked against your body with its feet up. Press your fingers firmly but gently on both sides of its genital area to push the genitals out. You will see a genital slit (for females) or the cylindrical shape of a penis (for males). On young rabbits, telling the difference can be tricky, even for professional eyes.

Is It Worth Buying Pet Insurance?

Pet insurance can be tricky. It all boils down to these questions: What choices are available in your area? What do the insurance companies offer? And how much do the policies cost? There are several options for pet insurance for dogs and cats, but it's rarer to find a pet insurance company that covers rabbits. It's important to read through the entire policy to check for any treatments or procedures the company does not cover. Most people just sign up without realizing their policies don't cover the most common issues rabbits can have. Make sure there are no exclusions for gastrointestinal (GI) stasis or liver torsion surgeries, the most common surgeries for rabbits. It's very frustrating to go to the emergency veterinary clinic with a common issue like GI stasis, only to learn it's not covered *because* it's so common.

Don't be afraid to ask questions before signing up. Pet insurance is typically not too expensive, but don't make the investment if you don't think it will be helpful.

Some male rabbits have split penises and can look like females when they are young—then, lo and behold, a few months go by and testicles start to form. It's good to have someone with a well-trained eye check the sexes of your rabbits, especially if they are young.

Common Health Issues

It's important to monitor your rabbit's health closely and seek veterinary care if you notice any signs of illness or irregular behavior. A yearly checkup is a good way to prevent serious health issues. Of course, if anything happens in between yearly appointments, don't wait. Get your bunny checked out as soon as possible. Rabbits can exhibit very similar behaviors for various illnesses, so it's good to establish care with an exotic-animal veterinarian in case of emergencies.

Here are some of the main health issues found in rabbits.

Digestive Problems

Gastrointestinal (GI) stasis is extremely common in rabbits and is the number-one cause of death. Rabbits have very sensitive GI tracts. An unbalanced diet, a stressful living situation, or dehydration can cause a rabbit's digestive system to back up to the point that the rabbit can't poop. This type of blockage can cause a lot of internal damage, and the rabbit can die within 24 hours if not properly treated. Signs and symptoms include hunching over, breathing slowly, shutting the eyes halfway or fully, and grinding teeth. A rabbit with GI stasis may also show no reaction when you push against it.

A rabbit owner should have a high-fiber recovery food on hand at all times. This is a powder made from high-fiber natural ingredients, which can help push everything through the GI tract. It can be administered at the beginning stages of GI stasis, if your rabbit is still moving around but has lost interest in food or hasn't pooped in several hours. Follow the directions and be careful not to give too much, because that can back up the rabbit's system even more.

If your rabbit is already immobile, in the later stages of GI stasis, you'll need an emergency trip to the veterinarian to save its life. In the worst cases, surgery may be needed to remove some of the blockage. GI stasis isn't a disease and isn't genetic; all rabbits are vulnerable to it. Reduce the chance that it develops by making sure your rabbit always has fresh water and hay and by avoiding high-carbohydrate treats and feeds. Ensure your bunny has room to run around and exercise daily, as this will help keep its GI tract moving.

Rabbit Hemorrhagic Disease

Rabbit hemorrhagic disease virus (RHDV) spreads easily and comes from wild rabbits. When this disease hits, it acts very fast. Possible symptoms include lethargy and blood around the nose from internal bleeding; once these symptoms appear, death can follow quickly. RHDV is spread through direct contact, so if a wild rabbit passed away and you stepped where its infected body had been, you can carry the virus on your clothes or shoes into your home. RHDV has been found to survive on surfaces for more than four months. It is usually fatal in rabbits; only a small percentage of those who get it survive. A second strain, RHDV2, is also lethal. Symptoms are similar to those of the first strain, and rabbits that do survive it can have neurological, respiratory, or gastrointestinal issues. There is a vaccine now available in most countries, so if your rabbit is at risk of exposure, you can keep it safe. Ask your veterinarian if RHDV is prevalent in your area and go from there.

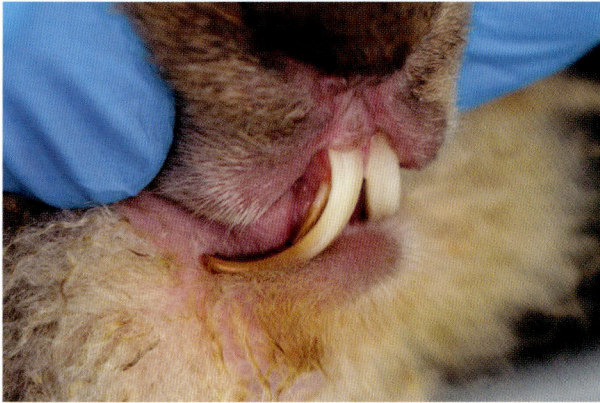

Malocclusion, a hereditary defect, causes misaligned teeth that need regular attention.

Dental Issues

Dental problems are common in rabbits. Since their teeth never stop growing, they need to wear them down by chewing. If the teeth don't wear down evenly, rabbits can suffer from abscesses, overgrown teeth, or malocclusion (misaligned teeth).

An abscess can be surgically removed, but surgery is risky for rabbits because they are very sensitive to anesthesia. Because a bad abscess is potentially life threatening, surgery may be the only option.

Malocclusion is hereditary, which is why breeders must carefully select their breeding stock and customers must ask about the dental health of a rabbit. Malocclusion is not fixable, but it can be managed to make the rabbit more comfortable and healthier. You can visit the veterinary clinic every few months to have your rabbit's teeth ground down with an electric rasp, a quick and harmless procedure. If the molars are misaligned, your bunny will be put under for safety, but that usually is not necessary for the front teeth, and you can even learn to do it yourself. Have a professional teach you before attempting it on your own.

Respiratory Ailments

Rabbits are also prone to respiratory infections, which can be caused by viruses, bacteria, or environmental factors such as poor ventilation or dirty living conditions. This is why it's very important to keep their area clean! If your rabbit gets a respiratory infection, make sure you clean its living space thoroughly. Bleach is very strong, so I like to use vinegar and baking soda instead. Meanwhile, a quick trip to the veterinarian and a round of antibiotics should help clear up the issue.

Pasteurellosis is an uncommon bacterial infection that affects the respiratory system and lungs of rabbits. Treatment through antibiotic therapy can be challenging and may take a long time. Symptoms can include difficulty breathing, discharge from the nose, red or watery eyes, sneezing or snoring, and swelling around the face.

Parasites

Parasites are a common occurrence in rabbits but are usually treatable.

Coccidiosis is a parasitic disease of the intestinal tract caused by coccidian protozoa. The disease can cause diarrhea, weight loss, lethargy, and even death. Adult rabbits with coccidiosis usually don't suffer any issues from it, but baby bunnies are very fragile and can easily die. Coccidiosis is usually spread through fecal matter. Sometimes mama buns carry it without any symptoms, but when baby bunnies eat their mom's droppings, as they normally do, the infection works quickly to make them very weak and ill. It's another reason to keep those cages clean! Parasites thrive in dirty environments.

Encephalitozoon cuniculi is a very common microsporidian fungal infection: More than 80 percent of rabbits have been exposed to it. Most rabbits with *E. cuniculi* live normal lives with no symptoms. A healthy rabbit with a strong immune system will most likely never have issues with this fungus, but it can potentially cause complications or life-threatening symptoms. *E. cuniculi* can affect the nervous system, leading to symptoms such as head tilt, seizures, paralysis, or kidney failure. (Head tilt doesn't always mean *E. cuniculi*. Sometimes a rabbit's head can tilt because of an ear infection, which is especially common in lop-eared rabbits, so it's best to have your veterinarian determine the cause.) Despite many years of research, there is no cure for *E. cuniculi*, but there are effective treatments. If you are wondering if humans can contract these parasites from rabbits, the answer is no. These parasites are species-specific.

Ear mites are common external parasites that cause irritation, itching, and discomfort. These mites are different from the skin mites found on other parts of a rabbit's body. They can be treated with medication, so if you notice your rabbit scratching excessively, take it in for a checkup.

Ear infections are most common in lop-eared rabbits. It can be hard to detect an infection. If you notice excessive scratching of the ears, you can treat them with a cleaning solution and see if that helps over time. If not, a veterinary visit might be warranted to get a diagnosis and the proper treatment.

Skin mites aren't usually a major concern, but rabbits can get mites that will eventually leave little bald patches, typically behind the neck. If your rabbit is scratching constantly and you notice patches of fur disappearing, flaky skin, or crusty scales, it could have skin mites. A veterinarian can prescribe an antiparasitic treatment that may need to be repeated until the mites are gone. It's also a good idea to deep-clean your rabbit's living area and wash any washable items with very hot water.

Ringworm is a fungal disease that shows up as circular red rings on bare spots of skin. It's most commonly seen in young rabbits, older rabbits, or rabbits with weak immune systems. Ringworm is easily treatable with topical medication or, if it's spread beyond a limited area of the body, with an oral medication. Once treatment starts, thoroughly clean the rabbit's living area.

Flystrike, or myiasis, is when flies lay eggs on your rabbit's fur or skin. The maggots that hatch out of these eggs feed on flesh, which can cause tissue damage and infection. This condition needs to be treated immediately. Flystrike is more common in the summer, when it's warmer and flies are more prevalent. If your rabbit is healthy and has no open wounds or urine soaking its underside, the risk is lower. Dirty conditions provide breeding grounds for flies. If flystrike occurs, your rabbit will need to be thoroughly cleaned by a veterinarian; anesthesia may be necessary if dead tissue needs to be removed.

Most ailments are readily treatable if caught early. Check your bunny regularly for signs of parasites or other issues.

Sore Hocks

Sore hocks, also known as pododermatitis, occur when your rabbit spends most of its time on hard floors or wire flooring. The back feet will start to lose hair, exposing skin to the hard surfaces and causing sores to break out on the feet, which can get infected. Rex-haired rabbits are more prone to sore hocks because they do not have as much fluffy hair on the bottoms of their feet as other rabbits do.

If you catch sore hocks in the early stages, not yet inflamed or infected, you can treat them with diaper rash cream or a medication prescribed by your veterinarian. Either one should clear the issue quickly and get your bunny on the road to recovery. If your rabbit is prone to sore hocks, you may want to add softer flooring or mats to its area.

Urinary Tract Problems

Rabbits can also suffer from urinary tract infections (UTIs), which may lead to bladder sludge or bladder stones. A high-calcium diet with foods such as alfalfa hay or alfalfa-based pellets, or even too much spinach, can cause kidney stones. It's important to always have fresh water available, as dehydration can cause UT issues. *E. cuniculi* and obesity are other causes of UTIs. Symptoms may include straining to urinate, urinating outside the litter box, urinating every time they are held, or frequent urination. Your veterinarian can prescribe medication to get rid of the infection.

Less Common Health Issues

It's important to know the signs and symptoms of uncommon issues as well. If you spot these, get to a veterinarian as soon as possible so that your rabbit's ailments don't continue or worsen.

Because GI stasis is so common, veterinarians once tended to diagnose it right away when they noticed its typical symptoms. But over the last decade, researchers have found that GI stasis may not be the only thing that can cause a rabbit to act lethargic, stop eating or pooping, or become hunched over and unresponsive.

Liver torsion, also known as hepatic torsion, is becoming more widely recognized in lop-eared rabbits. In this condition, the liver twists around its blood vessels, restricting blood flow and oxygen supply to the liver tissue. Rabbits suffering from liver torsion show the same signs and symptoms as those with GI stasis, but the treatment is a costly surgery, ranging from about $7,000 to $10,000. The surgery is effective but risky; untreated, however, liver torsion can lead to organ failure and eventually death. Though liver torsion is uncommon, it's diagnosed more often now that it is better understood.

Treponematosis, a version of syphilis, can infect rabbits. This is a different disease from human syphilis and cannot be transmitted to humans, but rabbits can carry and transmit their version of syphilis to one another with no symptoms. Once you have noticed an outbreak, all rabbits who have been exposed to one another should be treated. If a pregnant mama bun is carrying rabbit syphilis, her babies will be exposed to it as they are birthed. If your rabbit gets small blisters on its genital area and/or nose, or slow-healing sores that eventually are covered with heavy scabs, it may have rabbit syphilis. The disease is easily treatable with a series of penicillin injections. Most veterinarians will give the first shot and provide instructions for you to give the rest of the shots, but if you're too uncomfortable, you can pay them to do the others. If you do give the shots, they must be administered subcutaneously, meaning right under the skin. Penicillin should not be given orally to rabbits or injected intramuscularly or intravenously, as any of those methods can be fatal. Syphilis clears quickly and is not a medical emergency.

Heart disease in rabbits is unusual. The symptoms include lethargy and difficulty breathing.

Lymphoma, or cancer of the lymphatic system, can occur in rabbits but it's not very common. As with liver torsion, the symptoms are similar to those of GI stasis. Your rabbit may experience weight loss, lethargy, decreased appetite, and swollen lymph nodes.

In spite of the ailments discussed in this chapter, most rabbits live happy, healthy lives with few issues!

Index

Page numbers in *italics* indicate photos.

J

Jersey Wooly, 122

K

kit (baby bunny), 20

L

large breeds, 28. *See also* specific breed
lifespan, rabbit, 17, 21
Lionhead, 28, 38, *38*, 122
litter, 20, 112, *112*
litter box, 11, 22. *See also* potty training
    baby bunnies and, 22, 24
    cleaning, 15, 115
    gridded, 111, *111*
    litter options and, 111–12, *111*, *112*
    second bunny, adding, 83–84
liver torsion, 136
loaf/"loafed" rabbit, 20, 65, *65*
    distress signs and, 129
logomorph, 20
lymphoma, 137

M

male rabbit. *See* buck
malocclusion, teeth, 133, *133*
Mini Lops, 123
Mini Plush Lop, *27*
Mini Rex, *27*, 28, 39, *39*
molting, 123
more than two rabbits, 85–86, *85*
    third bunny introduction, 86
mounting behavior, 81, 83, *83*

N

nail trimming, 121, 125, *125*
    overdue for, 128, *128*
nest, 20

Netherland Dwarf, 28, 40, *41*, *42*, 123
neutering, 17, 129–131
    age for, typical, 22, 130
    fertility following, 26
    hormonal levels and, 24
    territorial behavior and, 26
new companion, introducing, 27
nipping behavior, 78
nose/nostrils, 56, *56*
    anatomy and, 18
    body language and, 64
    checklist, healthy, 127

O

older rabbits. *See* adult rabbits
opposite sex rabbit pairs, 25–26, 80
other pets, introducing rabbits to, 87–89
    cats and rabbits, 89
    dogs and rabbits, 87–89, *88*
outdoor playtime, 110
outdoors, rabbits living, 116–19, *116*
    hutches and, 118–19, *119*
    keeping cool and, 117
    not on grass or dirt, 118
    sheds and, 117

P

pair of rabbits, *80*
    adopting, 25–26
    dominant member, 80
    opposite sexes and, 25–26, 80
parasites, 134, 135
pasteurellosis, 134
pelleted food, 96–97, *96*, *97*
pens, indoor rabbits, 108–9
    acrylic panel, 108, *108*
    cleaning, 115
    flooring under, 109, *109*
    size of pen, 16

pet insurance, 17, 49, 131
pets, rabbits as great, 11, 13, 28
pet stores, avoiding, 50–51
picking up socialized rabbit, 71–73, *72*, *73*
playing behavior, 67, *67*, 69
playtime, outdoor, 110
pododermatitis, 135–36
poop, bunny, 61. *See also* cecotropes; droppings, rabbit
pooping behavior
    dirty cage and, 115
    new environment and, 59
    outside litter box, 114
poses/positions, 65–66, *65*, *66*
potty training, 113–14, *113*
    outside the box, 114
    urine-stained fur and, 124, *124*
purebred rabbits, 17

R

rabbit hemorrhagic disease virus (RHDV), 132
refunds/return clause, 48–49
rehoming older rabbits, *23*
respiration, 128
    distress signs and, 129
respiratory ailments, 133–34
Rex, *27*. *See also* Mini Rex
rex fur, *27*, 123, 135
RHDV (rabbit hemorrhagic disease virus), 132
ringworm, 135

S

same-sex rabbit pairs, 25
scammers, avoiding, 49
second bunny, adding, 80–82, *80*
    aggressive behavior and, 81, 84
    barrier between, 80, *80*
    bonding and, 80, 81
    introducing newcomer, 82–84, *83*

second bunny, adding, *continued*

> older rabbit and, 82
> pen setups and, 83
> veterinarian trips and, 84
> younger, the better, 81–82

setup, basic, 14, *14*

sex of rabbit, checking for,
  130–31, *130*

sheds, 117

shelters. *See* adopting from
  shelters

single rabbit, 25

skin, checklist for, 128

skin mites, 135

skittish rabbit, socializing,
  74–75, *74*

sleeping positions, 66, *66*

smaller breeds, 28. *See also* specific breed

socialization. *See also*
  dominance/dominant
  bonding and, 21, *21*
  breeders' rabbits and, 48
  handling and, 28
  time spent with rabbit, 25

socialized rabbit, handling,
  71–73, *72, 73*

socializing skittish rabbit,
  74–75, *74*

social media accounts, fake, 49

sore hocks, 128, 135–36

spaying, 17, 129–131
  age for, typical, 22, 130
  hormonal levels and, 24
  territorial behavior and, 26

spraying urine, 24, 77

stress, biting and, 79

supplies
  equipment and, 55, *55, 123*
  where to buy, 114–15

T

tail, anatomy and, 19, *19*

teeth, 57, *57*
  dental issues, 133, *133*
  grinding, 129

temperature, tail raising and, 19

terms, common rabbit, 20

territorial behavior, 77, 106
  adopted pair and, 25
  marking territory, 22, 77, 114
  spaying, neutering and, 26
  spraying urine, 24, 77
  time alone and, 82

third bunny introduction, 86

thumping, 20

toys, simple, 55, *55*

transition to home, 59–60

travel carrier. *See* carrier, travel

treponematosis, 137

U

urinary tract infections (UTIs),
  136

urine, spraying, 22, 77, 106, 114

urine-stained fur, cleaning, 124,
  124

V

vegetables. *See* greens

Velveteen Lop, *27*

vent area, healthy rabbit, 56

veterinarian, 58–59
  both bunnies, bringing, 84

vibrissae. *See* whiskers

W

warren, 20

water, 101–3, *102*
  bowl or water bottle, 102–3

weight of rabbit, 128

whiskers, 19

wild DNA, pet bunny and, 70, *70*